This book is a great resource for those of us communicating with funding agencies and preparing grant submissions to advance our work. Written by an insider, it is perfect for those new to the process as well as helpful to experienced applicants navigating rapidly changing rules and guidelines. I recommend it as a perfect addition for all scientists, regardless of area of research or funding source.

—**KARLENE BALL, PhD,** UNIVERSITY PROFESSOR AND ENDOWED CHAIR, UNIVERSITY OF ALABAMA AT BIRMINGHAM; FELLOW, AMERICAN PSYCHOLOGICAL ASSOCIATION

It isn't often one can find a book on grant funding that provides the broader context affecting the grant preparation process and the grant policy issues that may impact study design. Jeff Elias draws on his vast and diverse experience with research and grant funding to make the whole grants process more understandable and less daunting. As new grant opportunities continue to proliferate, this readable book will quickly become a go-to resource for novice and advanced researchers alike as well as members of the grant support workforce at universities. I highly recommend it.

—**GEORGE W. REBOK, PhD,** PROFESSOR EMERITUS, JOHNS HOPKINS UNIVERSITY, BALTIMORE, MD; FELLOW, AMERICAN PSYCHOLOGICAL ASSOCIATION

GET
FUNDED

GET

FUNDED

A PRACTICAL GUIDE TO UNDERSTANDING THE GRANT APPLICATION PROCESS AND WRITING WINNING PROPOSALS IN THE BEHAVIORAL AND BIOMEDICAL FIELDS

Jeffrey Wayne Elias PhD

 AMERICAN PSYCHOLOGICAL ASSOCIATION

Published by
American Psychological Association
750 First Street, NE
Washington, DC 20002
https://www.apa.org

Order Department
https://www.apa.org/pubs/books
order@apa.org

Typeset in Charter and Interstate by Circle Graphics, Inc., Reisterstown, MD

Printer: Lake Book Manufacturing, LLC, Melrose Park, IL
Cover Designer: Mark Karis

Library of Congress Cataloging-in-Publication Data

CIP data has been applied for.
Library of Congress Control Number: 2023935471

https://doi.org/10.1037/0000390-000

Printed in the United States of America

10 9 8 7 6 5 4 3 2 1

Contents

GET
FUNDED

INTRODUCTION

Who Is This Book For, and What Is It About?

Science is not static; it continues to evolve. Consequently, regardless of the funding experience and previous success of investigators and research institutions, the nature of funding continues to change. There are always new challenges and challengers. This book provides tactics and strategies to systematically prepare for the competitive funding process, including self and team evaluation, designed to promote retrospection, introspection, and constructive future expectations. I will explain how to expertly identify and convey what makes your proposed research impactful, innovative, and achievable.

Good ideas drive science forward, but funding agencies do not award funding solely on the basis of how good an idea is. Grant writing is a communicative endeavor. There is a broader context that affects grant preparation processes. Understanding and communicating within that broader context can enable a more positive application review. I will explain how and when to interact with funding institutions and how to interpret and respond to peer review feedback.

The chapters in this book provide a broad basis for interacting with funding agencies, most of which follow standard guidelines for submission, review, feedback, and funding. Readers are encouraged to apply the principles of

https://doi.org/10.1037/0000390-001
Get Funded: A Practical Guide to Understanding the Grant Application Process and Writing Winning Proposals in the Behavioral and Biomedical Fields, by J. W. Elias

grant writing outlined in this book to the specific funding policies and dead-liness that will characterize grant writing for any funding entity.

The suggestions in the text apply to the development of any funding oppor-tunity, including opportunities in the basic sciences. The strategies and tactics I discuss are most frequently examined in the context of applying to the National Institutes of Health (NIH) and the National Science Foundation (NSF). The size and breadth of research funding by NIH and NSF provide the most usable examples of how to navigate granting agencies more broadly. All funding agencies are mission driven in some fashion, and applicants need to understand the greater mission of a funding source before applying. NIH and the NSF exemplify how the social needs of society interact with basic science to produce products that are designed to move both science and society forward.

Communication with funding agencies and their agents is a two-way process. A primary source of communication for funding agencies is their website. Much of grant writing is knowing where to find the latest informa-tion on funding sources and then navigating a website. This text will direct readers to URL-based sources of information, which is sometimes like opening a book at the beginning only to find yourself in the middle chapters and not knowing how you arrived there. These are the portals to most funding sources, although many funding agencies provide manuals to explain their processes; these manuals can be downloaded by locating and opening the appropriate URL. Once you have located important content on a website, copy it and make it active on your list of grant resources!

The content in this book will be helpful for those just becoming initiated into the world of grant funding as well as for those who are more advanced. New researchers need to know how to identify a promising research topic, assemble a strong research team, and complete an application. This book will also help more advanced researchers, who may have multiple roles to play in the grant funding process. Advanced researchers are frequently called on to mentor others while maintaining their own funding. They also frequently work in administrative roles in the institutions that support them. Such dual roles require multiple perspectives on funding, which this book covers.

Most research institutions have a large administrative support workforce whose sole purpose is to help faculty apply for grants to support their research. These staff also need a broad understanding of the grants process and the factors that influence whether an application is funded. Thus, an additional audience for this book includes members of the grant support workforce at universities.

ABOUT ME

You will see certain themes develop throughout the text that are the result of my experiences with research and career development, and they represent the experiences of others as well. My first experience with funding and how it supports research and education came when I was an undergraduate in a program that received a federally funded grant to support the development of space and administrative resources to train undergraduates to develop real research projects with real mentors. I was supported through graduate school on research grants to my advisors and mentors, including a Public Health Services Fellowship. Using data from my dissertation, I submitted an application for a postdoctoral fellowship. It was funded, but I withdrew from the fellowship prior to funding and accepted a full-time academic position at a university. I expanded the postdoctoral fellowship application to an R01 application that was funded by the National Institute on Aging (NIA), which gave me a good start to a career in academics. That R01 included premise-supporting data and a couple of years of preparation.

My career experience with research has been purposely diversified and always involved research funding. This includes my experiences as an undergraduate student, graduate student, faculty member, research director, administrator, author, journal reviewer, journal editor, ad hoc and permanent peer review panel member, NIH Center for Scientific Review Officer, National Institutes on Aging Program Officer, and Director/Manager of Grants Facilitation at a school of medicine.

What I have observed from all these perspectives on research and funding is that there is a winning combination that includes access to space, a supportive environment, willing mentors, colleagues, student colleagues, research projects, support for data collection and student development, opportunities to diversify research and career interests, and time and resources to regroup and try again when things do not go your way. The resources of time and the ability of the support environment to adapt to grant funding policy changes for human subjects, clinical trials, and data management are becoming bigger parts of the winning combination.

Moving from the position of reviewer and applicant to the position of grant administrator at NIH was like discovering a new floor in my house. Grant review is much more complex and fast moving than the well-defined and limited environment of a study section makes it seem. Program Officers have a role that is quite broad and involves not only managing the best science possible but also promoting the advancement of that science, and the investigators, to include new investigators. The training role of administrative positions never ends because science and policy continually evolve.

OVERVIEW OF THIS BOOK

Chapters 1 through 3 of this book provide foundational information on the contexts of grant writing, including navigating the funding announcement and understanding the influence of grant policy issues on study design. Chapters 4 through 6 provide guidance for writing the different sections of a grant proposal. Chapters 7 and 8 provide an inside look at how peer reviewers from the funding institution evaluate proposals and how applicants can interpret and respond to peer review feedback. Chapters 9 and 10 focus on nontraditional grants, including training grants and product commercialization grants. Chapters 11 and 12 discuss how to advance grant writing skills by using team and time management and then moving toward advanced grant literacy, which ultimately can be defined as "knowing what to do next."

As grant writing and funding have progressed over the years, so too have the pomp and presentation of researchers and grant funding. This can be a bit daunting for those new to the competition, but, cheerleading aside, competition in grant writing requires a good game plan. For those newly initiated into the competition, I would say, "Why not you?"

PART **I**   **UNDERSTANDING
THE CONTEXT OF
GRANT FUNDING AND
PREPARING TO WRITE**

1 DEVELOPING GRANT LITERACY

In order to be successful at getting grants, researchers must develop what I call *grant literacy*. The use of the term *literacy* in the context of grant writing has many parallels to the use of the term *literacy* in health literacy. Title V of the Patient Protection and Affordable Care Act of 2010 (Pub. L. 111-148) defines *health literacy* as "the degree to which an individual has the capacity to obtain, communicate, process, and understand basic health information and services to make appropriate health decisions" (Office of Disease Prevention and Promotion, 2019, paragraph 3). Borrowing heavily from the health care definition of literacy (Elias, 2009), I define grant literacy as the degree to which an individual has the capacity to obtain, communicate, process, and understand basic grant information and services to make appropriate grant decisions.

The ability to navigate and recognize the multiple contexts in which applicants/investigators develop grant proposals and respond to feedback requires grant literacy. Researchers who embrace the goal of increasing grant literacy will not only have a better understanding of the granting process but also make better decisions. This book aims to increase the reader's grant literacy.

https://doi.org/10.1037/0000390-002
Get Funded: A Practical Guide to Understanding the Grant Application Process and Writing Winning Proposals in the Behavioral and Biomedical Fields, by J. W. Elias

One of the best ways to organize information regarding grant writing and funding is to examine the multiple contexts that drive the granting process. This chapter discusses seven key contexts that researchers should know about when applying for grants: (a) communication and access, (b) language and nomenclature, (c) funding agency contacts, (d) cycles of activity, (e) organizational systems, (f) competition, and (g) grant and funding policy.

COMMUNICATION AND ACCESS: WHERE CAN YOU FIND INFORMATION ABOUT GRANTS ONLINE?

Granting sources will communicate initially with applicants via websites that will explain how the granting process works for that funding source. Access to the websites is gained through the use of uniform resource locators (aka URLs). URLs are very useful and allow quick access to information, but sometimes the links provided are broken, or they contain old information. Websites provide immediate access to information, but over time they can become unwieldy as the funding process and the funding agency programs become more varied and more policy driven. The information on the funding websites can change quickly and is fluid. Major policy changes are usually announced well ahead of policy implementation. Applicants will have to adjust to the "too much information" and "too little information" nature of funding source websites.

These are some commonly used online sources for grant funding information:

- **Grants.Gov (https://www.grants.gov).** This website describes funding opportunities offered by the 26 agencies of the U.S. government. Although agencies also have their own websites that provide grant information and funding announcements, Grants.Gov has all of the funding opportunity announcements (FOAs). In 2023, federal funding agencies, including the National Institutes of Health (NIH), adopted the acronym NOFO to indicate a Notice of Funding Opportunity. These terms will likely both appear on FOAs until NOFO becomes, over time, the predominant acronym. These announcements all provide the needed information to apply for the funding, including the proper forms. Each agency has different terminology to refer to similar constructs or activities.

- **The National Institutes of Health (NIH) Office of Extramural Research website (https://grants.nih.gov/aboutoer/intro2oer.htm).** NIH is the largest funding agency for biomedical and behavioral research, with

27 research institutes and centers. According to NIH (n.d.), it has a $45 billion budget, and "over 84 percent of NIH's funding is awarded for extramural research, largely through almost 50,000 competitive grants to more than 300,000 researchers at more than 2,500 universities, medical schools, and other research institutions in every state."

- **The National Science Foundation (NSF) Proposal and Award Policies and Procedure Guide (https://www.nsf.gov/pubs/policydocs/ pappg22_1/nsf22_1.pdf).** NSF is an independent federal agency created by Congress in 1950. It supports basic science and engineering and has eight directorates: (a) Biological Sciences; (b) Computer and Information Science and Engineering; (c) Education and Human Resources; (d) Engineering; (e) Geosciences; (f) Mathematical and Physical Sciences; (g) Social, Behavioral and Economic Sciences; and (h) Technology, Innovation, and Partnerships. On its website (which is in rebuilding mode as of 2023), NSF describes itself as

  > [an agency] that propels the nation forward by advancing fundamental research in all fields of science and engineering. NSF supports research and people by providing facilities, instruments and funding to support their ingenuity and sustain the U.S. as a global leader in research and innovation. With a fiscal year 2023 budget of $9.5 billion, NSF funds reach all 50 states through grants to nearly 2,000 colleges, universities, and institutions. Each year, NSF receives more than 40,000 competitive proposals and makes about 11,000 new awards. Those awards include support for cooperative research with industry, Arctic and Antarctic research and operations, and U.S. participation in international scientific efforts. (https://www.nsf.gov/news/news_summ. jsp?cntn_id=296122&org=SES)

  Although the NSF general website is useful, the best source for understanding NSF programs is the NSF Proposal and Award Policies and Procedure Guide, published October 4, 2021, and available online (https://www. nsf.gov/pubs/policydocs/pappg22_1/nsf22_1.pdf). The 197-page manual is well organized, with a clear Table of Contents and page references that allow a specific topic to be quickly accessed if one is using the online manual. There is a small glossary of NSF-specific terms and general federal Grants.Gov terminology. See https://www.nsf.gov/bfa/dias/policy/ outreach/propprep_spring17.pdf for a good presentation of what NSF expects applicants to provide in a proposal. NSF was using NSF FastLane for submissions, but as of February 2023 it is using https://www.research. gov as a submission portal (https://www.nsf.gov/news/news_summ.jsp? cntn_id=305728).

- **The Department of Defense (DoD or DOD; https://www.defense.gov/).** DoD is a major resource for grant funding and funds a variety of science projects, including biomedical and behavioral ones if the proposed research is of interest to national defense and the military. DoD, for example, provides funding for cancer research and head injury research. The funding opportunities are referred to as *solicitations*, and they include Program Announcements; Broad Agency Announcements (BAAs); Requests for Proposals (RFPs); and Small Business Innovation Research (SBIR) grants, which are actually contracts and not grants in the sense of NIH SBIR grants. The solicitations allow for applicant-proposed solutions to DoD needs where outcomes are not required, a process that is similar to grants. DoD RFPs are contracts.

In addition to the websites just listed, you can do a general online search for the phrase "nonprofit organizations supporting research." This search will bring up additional sources for grant funding. Keep in mind, there will be broken links, so a careful search is required. Trust, but verify. Most universities and libraries can help with searching for nongovernmental funding sources. Many professional organizations provide resources for grant funding and research training.

The funding agency websites just listed are relatively stable with respect to information and links to information, but some of the links to information inside the websites might be broken. The websites have "search" components to accelerate the search for specific topics. Once you locate a URL for a specific topic, copy it and make it active on your own list of grant resources. If the link should become broken, go back to the original website URL. Bookmarking the URLs is useful, but again links can be broken or need to be updated. It is very important to try to tie the information you have requested to the date of that information.

LANGUAGE AND NOMENCLATURE: WHAT TERMS MUST YOU KNOW, AND HOW CAN YOU LEARN THEM?

Once grant resource websites have been located, individuals new to the granting process will immediately recognize that there is a nomenclature and language of communication that has to be mastered if one is to gain a full understanding of the information provided. NIH has such unique terminology that it has provided a glossary of terminology that is useful when applying for NIH funding (see https://grants.nih.gov/grants/glossary.htm). Some agencies will provide their own glossaries, which should be located within the website. NIH has also developed a separate glossary that applies to clinical trials (https://www.nih.gov/health-information/nih-clinical-research-trials-you/glossary-common-terms). The National Institute on Aging (NIA) has put together a similar clinical trial glossary that is quite useful (https://www.nia.nih.gov/research/dgcg/nia-glossary-clinical-research-terms).

There are important terms to define immediately. Most funding agencies distinguish among grants, contracts, and cooperative agreements. A *grant* is a means of transferring money, property, or services from a funding source (System 3 in Figure 1.1) to an institution (System 2 in Figure 1.1) that will provide on-site or proximal oversight for the applicants/investigators (System 1 in Figure 1.1) who will carry out the goals of the grant. A grant is not a loan, and it does not have to be paid back by the grantee, but the funds from a grant can be returned to the funding agency if the work on the grant is not carried out as proposed.

A *contract* requires delivery of specific contracted material—this could be data, an experiment, a program, or a material product. The contractor chooses a deadline for acquisition of materials and controls their distribution. Contracts can be created for science deliverables, but there is a focus on the services provided in the contract and a clear timetable for the services to be provided within the proposed budget.

A *cooperative agreement* is a service provided to a grantee or contractor that involves working closely with the funding agency to complete the project and monitor its progress. There is a category of grants offered by NIH called *U* (e.g., U01) grants that is referred to as a cooperative agreement because after the review process is complete and the grant is funded, the funding institute wishes to work in close concert with the grantees/investigators.

Other important terms to define immediately are *Program Announcement* (PA) and *FOA/NOFO*. A PA is defined in the NIH glossary as a "defined formal statement about a new or ongoing extramural activity or program. It may serve as a reminder of continuing interest in a research area, describe modifications in an activity or program, and/or invite applications for grant support" (https://grants.nih.gov/grants/policy/nihgps/html5/section_2/2.3.5_types_of_funding_opportunity_announcements__foas_.htm). *Extramural* refers to the funding offered to the public and non-NIH researchers who are supported

FIGURE 1.1. The Three-Systems Model of the Grant Funding Process

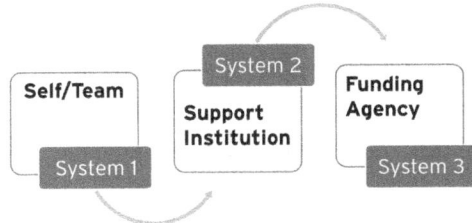

by an intramural funding program with its own source of funds and rules for funding. An FOA/NOFO is defined as

> a publicly available document by which a Federal Agency makes known its intentions to award discretionary grants or cooperative agreements, usually as a result of competition for funds. Funding Opportunity Announcements may be known as program announcements, requests for applications, notices of funding availability, solicitations, or other names depending on the Agency and type of program. (https://www.grants.gov/learn-grants/grant-terminology.html)

Both the PA and the FOA/NOFO are means by which to understand federal funding opportunities. This is an example of the tendency for federal agencies to use *binomial nomenclature*; that is, two terms are used to describe the same thing. Be aware that some funding sources use multiple coding schemes to distinguish between grant types. This use of multiple names can eventually be mastered, but it takes a little bit of experience to get used to the quirkiness. Instructions for navigating PAs and FOAs/NOFOs are discussed in detail in Chapter 2.

When reading NIH PAs or FOAs/NOFOs, a term applicants should become familiar with is *activity code*. These codes are used to designate among the many grant options offered by NIH. Not all of the 27 NIH institutes and centers fund all the grant activities. Therefore, when reading an NIH FOA/NOFO or PA, the applicant should pay attention to the NIH institutes that are signed onto it (i.e., will participate in the FOA/NOFO/PA). NIH officially defines a *mechanism* as one of three programs, referred to as (a) grants, (b) cooperative agreements, and (c) contracts, and occasionally refers to training grants as a grant mechanism. Three of the more popular activity codes are R03, R21, and R01. The R03 is small in terms of budget and time. The R21 is an exploratory grant that is restricted in budget and time and is designed to explore more risky aims. The R01 is the NIH's most common large research grant, offering a bigger budget and from 3 to 6 years of funding.

The multiple NIH grant activities distinguish among grants with different goals, funding levels, limits of funding, and renewable status. NSF refers to different grant activities in a much simpler way, as *programs*. A visit to the activity code website (https://grants.nih.gov/grants/funding/ac_search_results.htm) is helpful to gain an understanding how the activity codes are used in BAAs and grant programs.

NIH also provides an extensive list of acronyms that are used in referring to awarding entities, organizations, and other terms. A visit to the acronym list website (https://grants.nih.gov/grants/acronym_list.htm) will make clear how the list is useful for reading NIH PAs, FOAs/NOFOs, and other announcements, such as funding announcements.

The NSF, in much simpler fashion, refers to funding opportunities in terms of BAAs and grant programs. These are the gateways to the funding opportunities offered by NSF. For example, other funding agencies, such as the DOD, have their own nomenclature to discuss how solicitations are linked to mechanisms. A visit to https://www.universitylabpartners.org/blog/understanding-the-options-for-dod-funding illustrates how different funding agencies use and require knowledge of their own unique terminology.

FUNDING AGENCY CONTACTS: TO WHOM SHOULD YOU TALK?

Part of developing grant literacy is knowing whom to talk to at an agency and what to talk about. Much of this knowledge is developed through experience. Regardless of the source of funding, most FOAs/NOFOs will provide contact information. In the case of NIH, there are at least two, and sometimes three to four, contacts who have specific roles relative to the funding process (e.g., Scientific Review Officer [SRO],[1] Program Officer [PO], Referral Officer [RO], Grants Management Officer [GMO]). Some agencies, such as NSF, have a primary contact listed in the FOA/NOFO. A primary goal for funding applicants is to determine whom the key contacts are who can clarify information in an announcement. The use of agency contacts is discussed throughout this book.

CYCLES OF ACTIVITY: WHEN ARE THE DEADLINES?

A major context of grant funding is adjusting to a cycle of proposal deadlines and a cycle of waiting for funding. The cycles can differ in length for each funding agency, but in general there are periods of time that are set aside for (a) preparing a proposal; (b) the funding institute's processing of a proposal; (c) the funding institute's review of a proposal and preparation of feedback; and (d) if funding is awarded, there is the cycle of completing the proposed research—which follows the proposed timeline. If funding is not awarded, and a resubmission seems possible, then the first three cycles begin again as part of the resubmission process. For example, typical cycles of (NIH) funding are shown in Figure 1.2. There can be several months of time between the submission of an application and notification of the applicants'/investigators' receipt of funding or no funding. Almost all funding sources have cycles of function as opposed to open receipt and funding of grants.

---

[1]The NIH transitioned from the term Scientific Review Administrator (SRA) to Scientific Review Officer (SRO), but the SRA designation lives on and communications with the NIH may occasionally refer to the SRA.

FIGURE 1.2. National Institutes of Health Deadlines for Cycles of Submission, Review, and Funding

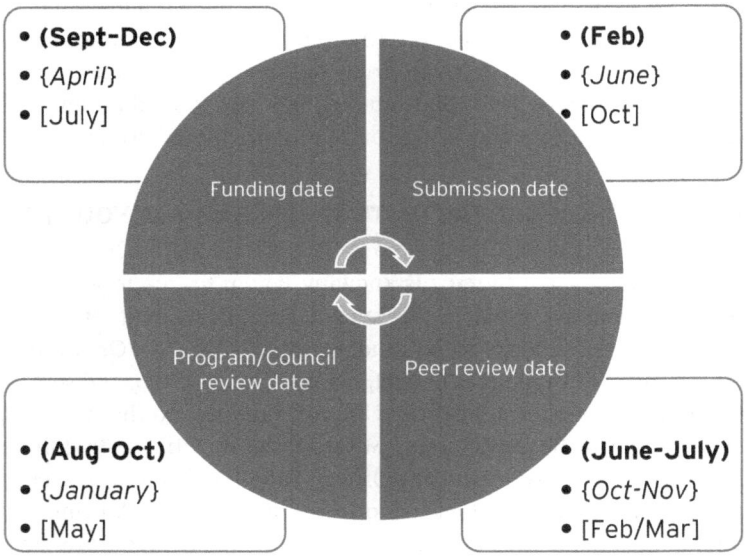

- **(Sept-Dec)**
- {*April*}
- [July]

Funding date

- **(Feb)**
- {*June*}
- [Oct]

Submission date

Program/Council review date

Peer review date

- **(Aug-Oct)**
- {*January*}
- [May]

- **(June-July)**
- {*Oct-Nov*}
- [Feb/Mar]

To manage the granting process, most funding resources permit submissions only at certain times of the year. Submissions occur for specific grant mechanisms within specific time periods. In other words, there are typically "hard deadlines," and grant applicants have to adjust not only to cycles of activity but also to the constraints of deadlines. Applicants who are submitting for multiple sources of funding may have to manage multiple deadlines. These cycles of activity have to fit in with other cycles of activity that academic, clinical, or research positions of employment require. One's personal time and personal life must fit within the same set of cycles.

ORGANIZATIONAL SYSTEMS: WHO IS INVOLVED IN THE GRANT FUNDING PROCESS?

Figure 1.1 introduces the three-systems model as an organizing principle for understanding the multiple functional contexts of grant writing and funding. The granting process is conceptualized as the functioning of three interacting systems of self and team, home (supporting) institution, and funding agency. This model follows the guidelines for business system models described by Bourgeois (2014).

This model not only helps organize the multiple operational contexts of grant writing but also provides a method of evaluating how each system is functioning relative to the goal of obtaining and supporting grant funding. Within the model, System 1 (self and team [i.e., applicant/investigator]) needs and uses research support from System 2 (home [supporting] institution) to develop proposals that can be funded by System 3 (funding agency). System 2 invests in System 1 to provide research dollars and recognition as a research institution. Research dollars and subsequent promotable research findings are used to build a reputation and attract students and scholars.

System 2 participants and administrators can acquire grant literacy, and eventually advanced grant literacy, by virtue of functional assessments of their ability to maintain the support context at a high level. Counting the number of grants funded to include center grants and training grants is one metric, but funding can also be centered within a minority of faculty, and that can produce pressured functioning with System 2. It is important that System 2 administrations quickly assess the impact of impending and existing System 3 policies as part of their support role for System 1. System 3 sets most of the policy related to grant funding, but System 3 would be expected to evaluate policy and make corrections, where needed, to provide System 1 and System 2 opportunities to develop grant literacy and, eventually, advanced grant literacy (see Chapter 10).

An overlooked but growing component of System 2 functioning and support is the grant support workforce (GSWF). The GSWF consists of individuals whose function is to assist with grant development as supported by System 2. The development of an educated GSWF is not quite an epiphenomenon but more of a by-product of the growing need for both System 2 and System 3 entities to develop a skilled group of individuals with a variety of skill training and interests, including math/statistics, writing/editing, data analysis and data management, cybersecurity, public relations, software and technology use, policy knowledge and application, personnel management, and social skills (aka *soft skills*).

Many in this workforce have received advanced degrees in science and related areas, and the GSWF network provides a good outlet for the application of their interests and training. Early training does not always dictate later career interest, but there is a special satisfaction in receiving science-related training and then applying it within a research environment, in particular when the training represents a sunk cost of effort and finance. The GSWF functions within the same work–environment contexts as the investigator workforce. For this specialized workforce there is the pressure of grant cycles, deciphering multiple funding opportunities, meeting multiple demands from multiple sources, and providing constant and/or last-minute reports on various aspects of grant funding and training.

The national effort to maintain the science (STEM) workforce through training and job opportunities has traditionally focused on the pipeline of potential authors and funded investigators (see https://grants.nih.gov/ngri.htm), that is, independently funded principal investigators. Despite the importance of the GSWF, other than job opportunity listings by research institutions that suggest needed skills and background training, little of a formal nature is known about how the national GSWF developed, how careers are advanced and maintained, or the degree of turnover and dropout. Nevertheless, a check of the job listings for major research institutions in particular likely will show several openings for people with the skills needed to maintain and grow a GSWF.

System 3 funding agencies award directly not to System 1 researchers but to the System 2 institution that is responsible for managing and distributing the funds to the System 1 applicants/investigators; that is, the funds from System 3 pass through System 2 for use by System 1. System 3 (funding agency) requires both System 1 (applicants) and System 2 (institutions) to provide research proposals, participate in peer review and fill stakeholder positions (e.g., council positions, special panels, visiting administrative positions), and conduct the funded research. System 3 funding consists of total dollars that consist of direct costs and indirect costs. The *direct costs* reflect the budget submitted with the application to cover salaries and the cost of the specific research project. The *indirect costs* reflect a negotiated percentage of the direct costs that System 3 funding institutes pay in facility support costs to the System 2 institution. The negotiated percentage differs between System 2 institutions on the basis of the cost of doing business in specific universities, medical schools, and geographic locations.

System 3 is most often funded by federal dollars, and the federal budget dollars are used to maintain System 3 functionality (e.g., staff, resources, buildings). The Congressional Research Service provides extensive reports on research funding trends and allocation of funds (https://sgp.fas.org/crs/misc/R46341.pdf). These reports are worth reading for all groups represented in the three-systems model because they provide an overall context for how congressional research and development is distributed to meet the needs of science.

System 3 supports research training grants and postdoctoral positions to provide a trained workforce to advance science and provide human capital support to all three systems. To keep the three-systems model functional, and to meet the national basic science, health, and biomedical research goals, it is incumbent upon each system to become the best functional version of itself. Individuals in System 1, the self/team system, must have the ability to

- assimilate information;

- navigate support systems and funding sources;

- effectively communicate;

- accommodate change;

- manage time and grant funding cycles;

- develop self and team skills;

- collect and publish data;

- motivate self and team members;

- adapt to technological platforms that shape how information is distributed between investigators, the home institution, and the funding institutions; and

- evaluate satisfaction with participation in the granting process and science.

Individuals in System 2, the supporting institution, must be able to

- provide physical resources, such as an office, research space, equipment, supplies, research assistants, and possible professional travel;

- prepare and submit grant budgets and eventually manage grant funding;

- provide access to colleagues, including the GSWF;

- provide access to pilot funding;

- provide the opportunity for research time to collect pilot data and prepare a proposal while receiving salary support from the institution;

- assimilate and disseminate grant-related information; and

- manage time and grant funding cycles.

Individuals in System 3, the funding agency, must be able to

- create, disseminate, and enforce multiple rules for submission, eligibility for submission, budget constraints and limits, and multiple forms for submission;

- follow multiple internal processing stages;

- communicate with multiple personnel regarding various issues;
- create and disseminate multiple announcements using specific language and terminology that must be learned;
- navigate multiple grant mechanisms;
- oversee policy that is constantly evolving, requiring adjustments to all three systems;
- manage funding and issue progress reports;
- respond to stakeholders; and
- establish cycles for grant preparation, submission, review, and funding (potentially resubmission).

The three-systems model is simple in inception and conception but complex in function. Becoming literate within the functional complexity of the three-systems model requires an understanding not only of the processes of grant submission (writing, submission, review) but also the context supporting the processes and the culture involved in grant funding.

It is important to note that any viable product produced in a federally funded grant could potentially be patented. All grant applicants should make themselves aware of the federal options to patent grant products by researching the Bayh–Dole Act (Pub. L. 96-517; also called the Patent and Trademark Law Amendments Act of 1980; see https://en.wikipedia.org/wiki/Bayh%E2% 80%93Dole_Act). An exception to the general System 3 model for developing funding for products is small-business entrepreneurial funding (https://www. sbir.gov/node/2100269; see also Chapter 9), aka SBIR/Small Business Technology Transfer (STTR) funding, This funding is designed to develop a product that can be patented (https://www.investopedia.com/terms/p/patent.asp) for property rights and potentially licensed. Investigators and institutions can patent products developed from non-SBIR/STTR grants, but the initial goal of the SBIR/STTR is to develop a product, find venture capital, and then market a product. As noted, any funded System 1 grant can develop a product that can be patented (see Bayh–Dole Act), but the SBIR/STTR application is specifically focused on the science of developing a product and eventually marketing the product.

Small-business grants include System 1 grant proposals, but small-business grant payments go to the company that may be collaborating with a funding institution, or to the funding institution that is supporting a company. Eventually, venture capital funding must be provided for the final product to be promoted and developed to its fullest extent. SBIR and STTR grants are more complicated than the basic System 3 model that fits most grant mechanisms. Small-business grant applications are discussed in Chapter 9.

Functions within the context of System 1, System 2, and System 3 are not static. The three systems change within their own context, but at varying rates. System 1 changes more rapidly in function because teams and investigators gain experience, change collaborations, and develop new skills and research interests. System 2 contexts change at a slower rate via changes in mission, financial resources, human capital and collaborations, and in administrative changes. System 3 changes occur more slowly but do so by virtue of trends in science, social issues, and policy that determine functioning and the distribution of financial resources. Sometimes changes in administrative leadership can influence the context of function in System 3 just as they affect the functioning of System 2. Functional context changes in System 3 are more likely to affect the context of function of System 2 and System 1 than vice versa. One feature of developing grant literacy as part of System 1 is understanding and knowing how to respond to immediate or pending changes in the functional context of System 3 and System 2.

COMPETITION: HOW CAN YOU STAND OUT FROM OTHER APPLICANTS?

For those involved in science, the dual components of competition and evaluation are introduced early in academic classrooms. Applying for a job frequently involves competing with others at some level. Individuals in research occupations quickly learn that the evaluation process involves a peer review process that starts with the attempt to publish an article. The peer review system in place for most federally funded grant applications is an extension of the normal competition and evaluation of science products. Peer review of grants is celebrated as a method of evaluation because of its impact on funding. Peer review is not where the competition begins; competition begins with how well System 1 and System 2 can manage their specific system responsibilities.

Grant applicants who have experience in the grant review process will find that achieving a good score can be related to how well the proposed ideas are presented. Smoothness and clarity of presentation in a document allows for assigned reviewers to better understand the proposed aims of a proposal. Clarity in a proposal allows reviewers to better organize the presentation of an application to a peer review panel.

Recently, the NIH Center for Scientific Review (CSR) proposed changes to the review process that the public can comment on (see https://public.csr.nih.gov/sites/default/files/2022-09/CSR-strategic-plan.pdf). The changes are not radical, but they will affect how proposals are reviewed and the scoring of each of the criteria listed earlier. These proposed changes are discussed further in Chapter 7.

Although peer reviewers do not officially score the degree of clarity and polish in an application, they may comment on these factors when they discuss the application with others on the review panel, and this may affect not only the final score of an assigned reviewer but also the scores of listening panel members. Thus, grant peer review is a bit of a beauty contest as well as a science contest.

GRANT AND FUNDING POLICY: WHAT ARE THE REQUIREMENTS FOR YOUR APPLICATION?

Readers will see in the ensuing chapters that policy drives the review and funding process for government funding agencies. By *policy*, I mean the rules and procedures that agencies, and the researchers they fund, must follow. Applicants cannot escape or ignore policy, even if a policy simply pertains to formatting a document. Investigators sometimes expect that policies with low personal face value can be ignored. This is not necessarily the case. Stated policies must be followed, even if they do not make sense or meet beliefs about best practice.

Investigators should remain alert for funding agency announcements of policy changes or increased policy enforcement. Funding agencies and investigators can become involved in public disagreements about the application of rules. A current example of this public disagreement is the NIH requirement that some basic experimental studies with humans (aka *BESH studies*) register as clinical trials for data management and reporting purposes (this is discussed in Chapter 3). Occasional messiness aside, understanding and following grant policy is part of acquiring grant literacy and accommodating the culture of competitive grant writing.

Investigators typically want to focus only on the writing and data analysis components of grants and try to address policy on a "just in time" basis. One of the most significant challenges to administrative support personnel is staying current with the language in policy announcements so that there can be accurate forecasting of the downstream outcomes of pending and current grant funding policies.

Funding agencies rarely introduce new policies without the benefit of comment periods that are open to the public, ample time to prepare for policy application, or both. When investigators are aware that grant-related policies are about to change, the need to respond to the changes should be factored into the timing of submission of a proposal. System 2 upper level administration in research institutions can be reluctant to respond to upcoming policy changes, in particular, those that pose a challenge to existing organization and resources. For example, federally driven changes in the definition of a clinical trial; trial management; clinical trial submissions; the definition of research with human subjects; and, most important, changes in data management and data sharing for clinical trials, are some of the more recent federal government policy changes that require an increase in both the training and development of unfunded resources within institutions.

For example, when NIH notified research institutions by announcements that it might levy fines and grant funding suspensions if clinical trial policies are ignored, many research institution clinical trial offices were created or expanded to ensure that researchers could remain competitive. The pressured rollout of the policies surrounding the clinical trial process, clinical trial reporting, and clinical trial research institutions is still affecting System 2 home institution resources. The System 3 policy changes were prompted by the overall efforts in science to increase the fidelity of funded data by revealing more about the conditions of data gathering and management and eventual data sharing.

System 1 applicants preparing their initial grant proposal are less likely to be focused on upcoming changes in policy and are more likely focused on existing policy. Nevertheless, any investigator hoping to develop a stream of funding or serve as a mentor to other investigators needs to be aware when future policy changes are proposed or announced. Funding institute policy announcements always allow a period of several months for comments from the public or System 1 and System applicants. Most funding agency policy issues and changes can be quickly understood and managed. Some policies are more difficult to implement and require more planning by Systems 1, 2, and 3.

KEY TAKEAWAYS FROM CHAPTER 1

- This chapter introduced the concept of grant literacy to indicate that knowledge of the funding process develops with experience and continues to evolve such that better decisions are made with advanced knowledge.

- Seven contexts are presented to illustrate the multicontextual nature of the funding process, including the context of competition.

- Each context requires the need to acquire information while at the same time adapting to and accommodating the fluid nature of information within each context.

- A systems-based context is introduced to illustrate that each applicant/ investigator is required to function simultaneously within the self and team, supporting institution, and funding source environment and that function within each environment can be evaluated.

- The ensuing chapters discuss the knowledge and competencies that can be used to perform well within the overarching context of funding.

ONLINE RESOURCES FOR CHAPTER 1

- Access to federal funding agencies and grant programs, https://www.grants. gov/

- NIH glossary, https://grants.nih.gov/grants/glossary.htm

- NIH acronym glossary, https://grants.nih.gov/grants/funding/ac_search_ results.htm

- NIH clinical trial glossary, https://www.nih.gov/health-information/nih-clinical-research-trials-you/glossary-common-terms

- List of NIH institutes and centers, https://www.nih.gov/institutes-nih/ list-institutes-centers

- NIH budgets, https://www.nih.gov/about-NIH/what-we-do/budget

- NIH Next Generation Researcher Initiative, https://grants.nih.gov/ngri.htm

- History of NSF development, https://www.nsf.gov/about/history/overview-50.jsp#1940s)

- Access to and descriptions of NSF directorates (profiles of the National Science Foundation Directorates), https://www.nsf.gov/statistics/directorate-profiles/

- NSF policies for grant submission and award manual, https://www.nsf. gov/pubs/policydocs/pappg20_1/nsf20_1.pdf

- Slide presentation from the NSF website (2017) that shows the elements of an NSF proposal, https://www.nsf.gov/bfa/dias/policy/outreach/propprep_spring17.pdf

- Options for DoD funding, https://www.universitylabpartners.org/blog/understanding-the-options-for-dod-funding

- Federal funding amounts and resources, https://sgp.fas.org/crs/misc/R46341.pdf

- Bayh–Dole Act, https://en.wikipedia.org/wiki/Bayh%E2%80%93Dole_Act

- Understanding the patent process, https://www.investopedia.com/terms/p/patent.asp

- Licensing versus patents, https://flossbok.org/licenses-vs-patents/

- How federal research budgets are allocated, https://sgp.fas.org/crs/misc/R46341.pdf

2 NAVIGATING A FUNDING ANNOUNCEMENT

Regardless of an applicant's (aka investigator's) degree or level of experience, they will have to explore a System 3 (funding agency) website for funding opportunities and submission instructions. Grant opportunity announcements often may be referred to as *funding opportunity announcements* (FOAs) or, as of 2023, *notices of funding opportunity* (NOFOs; terminology is discussed in more detail shortly) that are the gateway to understanding what applicants/investigators must provide for the next submission.

All the funding sources have websites and nomenclature describing the granting programs and specific opportunities. Chapter 1 provided a number of important URLs one can access to explore funding and funding opportunities within different U.S. agencies. Other countries will have similar URLs that point to sources of research grant funding. Often there are national restrictions on out-of-country submissions.

A list of foundations supporting research can usually be obtained from System 2 (home [supporting]) grant administration offices, or by searching the internet. For investigators new to a funding source, the FOA/NOFO provides insight into the processes of System 3 funding. The management roles

https://doi.org/10.1037/0000390-003
Get Funded: A Practical Guide to Understanding the Grant Application Process and Writing Winning Proposals in the Behavioral and Biomedical Fields, by J. W. Elias

of the three systems, which were introduced in Chapter 1, become evident as the FOAs/NOFOs reveal how each of the management components is engaged in the funding process.

FOAs/NOFOs ENGAGE WITH SYSTEM 1 MANAGEMENT STYLE

As noted in Chapter 1, the rules for System 1 (self and team) to interact with System 3 are determined by the System 3 funding source. Thus, System 1 has a motivated interest in learning the rules. For all levels of experience in grant writing, the FOA/NOFO engages the 10 management areas introduced in Chapter 1 and expanded in Exhibit 2.1. For those motivated to participate in the grant submission process, these 10 self-management elements will be automatically engaged and continue to develop as one gains continued experience in navigating System 2 and System 3.

In Chapter 1, competition and the cyclic processes of preparation, submission, processing, and review were introduced as primary contexts for grant funding. As noted in Chapter 1, it would be a mistake to overlook the fact that competition begins with the ability of System 1 and System 2 to develop and manage their respective System 1 and 2 responsibilities. One of the most important competitive responsibilities and skills is the management of time, in particular because time is related to the cycles of funding and deadlines noted in the FOAs/NOFOs. System 3 review is the final point of competition. Last-minute preparation as a management style can make System 1 applicants less competitive, which of course means System 2 is less competitive given that System 2 receives the funds for System 1 management.

OPPORTUNITIES FOR NEW APPLICANTS/INVESTIGATORS

Many funding agencies recognize that new applicants/investigators are in the learning stage, in particular with respect to managing the time element and the funding agencies may provide early support funding mechanisms (see Chapter 3). For example, the National Institutes of Health (NIH) glossary (https://grants.nih.gov/grants/glossary.htm) provides the definitions of both an Early Stage Investigator (ESI) and a New Investigator (NI), categories that have different statuses at NIH.

NIs have not previously secured R01 (basic large grant) or equivalent funding. An ESI is a Program Director/Principal Investigator (PD/PI) who has completed their terminal research degree or reached the end of their postgraduate clinical training, whichever date is later, within the past 10 years

EXHIBIT 2.1. Expanded System 1 Developmental and Management Responsibilities

1. Assimilate
   - terminology
   - communication styles
   - policies
   - forms and submission guidelines
   - opportunities and announcements
   - feedback
2. Navigate
   - the institute's organization
   - the institute's modes of communication
   - institute personnel
   - the home institution
3. Communicate with
   - funding sources
   - colleagues
   - the funding team
   - the home institution
4. Accommodate
   - changing policies and rules
   - time cycles
   - positive, mixed, and negative feedback
   - changing personnel
   - advances in science
   - changing support—either more or less
5. Time management
   - time for grant development
   - time devoted to nongrant professional activities
   - personal responsibility time (nonprofessional life)
6. Develop
   - research ideas
   - research skills
   - staff
   - research team
7. Collect and publish data
   - establish data sources
   - set up data repositories
   - write and submit articles for publication

(continues)

EXHIBIT 2.1. Expanded System 1 Developmental and Management Responsibilities (*Continued*)

8. Motivate
   - self
   - personnel and teams
   - ideas and methods
9. Adapt to distributed cognition, that is, information distributed across and shaped by technological platforms.
   - Systems 1, 2, and 3 communication processes
10. Evaluate your satisfaction with participating in the granting process and science.
    - satisfaction with the submission and funding processes

and who has not previously competed successfully as PD/PI for a substantial NIH independent research award. A list of NIH grants that a PD/PI can hold and still be considered an ESI can be found at https://grants.nih.gov/policy/early-investigators/list-smaller-grants.htm. ESIs should peruse the policy of any funding agency with which they are working to see whether, similar to NIH, it has ESI policies. The National Science Foundation (NSF) has its own detailed opportunities for NIs that focus on tenure-track positions. A quick way to investigate these programs is to search the FAQ section of the NSF site (https://www.nsf.gov/publications/pub_summ.jsp?ods_key=nsf22110).

KEY ELEMENTS OF A FUNDING ANNOUNCEMENT

Regardless of the funding source, when applicants/investigators read an FOA/NOFO for planning and management guidance, they should find there are 10 key elements revealed in it. These elements are described in Exhibit 2.2.

NIH has the most complicated nomenclature and the most complex glossary of terms. FOAs/NOFOs are a category of Program Announcements (PAs) and contain a great deal of information. The terminology in an announcement may require a glossary to understand. As noted in Chapter 1, struggling with the language and terminology of an FOA/NOFO is normal even for those experienced with grant writing. FOAs/NOFOs may seem oddly organized, but funding organizations rely on the structure of the announcement to provide key information. Funding sources often will provide a number to identify FOAs/NOFOs and PAs so they can be more easily referenced and tracked by the funding source and the applicants. The larger the funding agency, such as NIH, the more details are provided in the FOA/NOFO/PA. Once again, the

EXHIBIT 2.2. The 10 Key Elements of a Funding Announcement

1. Required focus of a grant proposal, indicating the area of science of interest to the funding source

2. Funding dollar and year limits for a proposal

3. Named grant category for that funding source (e.g., NIH R01, R34, R21, R03; determine page limits, funding limits, and renewability)

4. Organizations with multiple components may support only some of the science in a general announcement. For example, NIH has multiple institutes. Not all institutes support all FOAs/NOFO or grant types. NSF has multiple directorates.

5. Period for submission, review, and possible resubmission

6. Required format for the proposal (e.g., forms, specific sections, number of pages, spacing, font)

7. Clinical trial or no-clinical-trial option (significant for human subjects and data management). This is an important category for NIH.

8. Eligibility of the investigator and institution (i.e., who can submit a proposal)

9. Officials who can be communicated with (contacted) at the funding source about the proposal

10. Procedure for submitting the proposal; this may include a letter of intent or a white paper that evaluates the general project goals before being allowed to proceed with the submission.

websites for large funding agencies provide a portal through which individuals can access FOAs/NOFOs/PAs.

Some funding sources do not provide glossaries of terms, or provide only incomplete glossaries, so the investigator will have to make their own list. As noted, the NIH website provides an acronym and terms glossary to facilitate assimilation of terminology (https://grants.nih.gov/grants/glossary.htm).

Funding Opportunity Announcement (FOA)/Notice of Funding Opportunity (NOFO): A publicly available document by which a federal agency makes known its intentions to award discretionary grants or cooperative agreements, usually because of competition for funds. FOAs/NOFOs may be known as *Program Announcements*, *Requests for Applications*, *notices of funding availability*, *solicitations*, or other names, depending on the agency and type of program. FOAs/NOFOs can be found at https://www.grants.gov/ and in the NIH Guide for Grants and Contracts (https://grants.nih.gov/funding/searchguide/index.html#/). In addition, NIH and other agencies of the Department of Health and Human Services have developed omnibus Parent Announcements for common grant mechanisms that have transitioned to electronic submission, for use by applicants who wish to submit what were formerly termed *unsolicited* or *investigator-initiated* applications. Part of attaining grant literacy and self-management is acquiring the knowledge to navigate websites and the information on those websites.

NIH Program Announcement (PA): a notice that does the following:

- It identifies areas of increased priority and/or emphasis on particular funding mechanisms for a specific area of science.

- It provides information on standard receipt (postmarked) dates on an ongoing basis. Standard receipt dates can change. Always check for the most recent announcements provided by a funding source and for the specific program that is supported by the announcement. Submitting institutions have internal receipt dates based on their own System 2 functioning and staffing. Find out when an application has to be provided to the supporting System 2 institution.

- It remains active for 3 years from the date of release, unless the announcement indicates a specific expiration date or the NIH institute or center inactivates sooner (see https://grants.nih.gov/grants/guide/notice-files/NOT-OD-05-025.html for more information on expiration dates).

PAR: A PA with special receipt, referral, and/or review considerations, as described in the PAR announcement.

PAS: A PA that includes specific set-aside funds as described in the PAS announcement.

Request for Application (RFA): an announcement that

- identifies a more narrowly defined area for which one or more NIH institutes have set aside funds for awarding grants;

- usually has a single receipt date (received on or before) specified in the RFA announcement; and

- is usually reviewed by a Scientific Review Group convened by the award-issuing component.

Request for Proposal (RFP): an announcement that solicits contract proposals. An RFP usually has one receipt date, as specified in the RFP solicitation. A contract is different from a grant because it requires a specific set of determined deliverables within a specific period, and competing for the contract depends on convincing reviewers that the specified deliverables can be delivered within the specified time frame.

Notice of Special Interest (NOSI): a notice posted in the NIH Guide for Grants and Contracts that succinctly highlights a specific topic of interest, for example, a specific area of research or program. These notices direct applicants to one or more active FOAs/NOFOs (often Parent Announcements) for the submission of applications for the initiative described. NIH can issue NOSIs quickly without the full amount of information found in a PA or FOA/NOFO. Many PAs will be issued as NOSIs going forward.

Notice (NOT): an announcement of policy and procedures, changes to RFA or PA announcements, RFPs, and other general information items.

Inquiries about specific NOTs, PAs, and RFAs published in the NIH guide should be directed to the NIH staff member or members identified in each announcement. As noted, NOTs are very important because they follow the initial announcements with updates to the initial announcement (NOSI), and it is assumed that anyone using the initial announcement for some purpose is also following the additional NOTs. This is like checking the bulletin board for updates to general assignments to duty in the military. Working on a proposal guided by a specific NOSI, RFA, FOA/NOFO, or PA requires that one compulsively look for the notices. This advice applies to any funding source. Sometimes there will be specific publications of the notices on a website or, for NIH, in the NIH guide, or applicants can continue to place the URL for the original FOA/NOFO into a search engine to look for new notices attached to the original FOA/NOFO, for example, PA-20-185 (Parent R01 Clinical Trial). Readers of grant announcements should always check the dates of the announcements to see if they are current. RFAs and PAs have a stated expiration date and must be renewed. When these announcements are renewed, they will receive an updated number and issue date.

Numbering System: Examples From the NIH Website

PA numbering (e.g., PA-06-008): Indicates a PA issued in 2006 or for funding in 2006 (06) with an associated serial number (008)

RFA numbering (e.g., RFA-HL-06-004): Indicates an RFA issued by the National Heart, Lung, and Blood Institute (HL) in 2006 or for funding in 2006 (06) with an associated serial number (004)

NOT numbering (e.g., NOT-OD-06-025): Indicates an NOT issued by the Office of the Director (OD) in Fiscal Year 2006 (06) with an associated serial number (025)

Quite possibly there is more information in the FOA/NOFO than will be used immediately. Much of the "extra" information is in an FOA/NOFO if you need it for further information or feel the need to verify knowledge that was previously obtained. For example, a PA announcement such as PA-20-185. As shown in Exhibit 2.3, the information in a Parent Grant announcement can contain as many as 36 pages of information. The good news is the announcements have an organized schema, and NIH program announcements contain a Table of Contents. In PA-20-185, the Table of Contents appears on printed page 14. The Table of Contents appears at the end of Part 1 (overview information) and prior to Part 2 (full text of the announcement). Why does the Table of Contents appear at the end of Part 1 and not at or near the beginning of Part 1? Unknown!

Nevertheless, the very good news about PAs (and RFAs FOAs/NOFOs, and NOSIs) is that much of the information needed to put grant writing and grant-writing procedures into an organized format can be found by reading the FOAs/NOFOs (PAs, RFAs, NOSIs) while looking for the 10 key elements of an announcement listed in Exhibit 2.2.

FINDING INFORMATION IN A PARENT GRANT

For NIH, the term *Parent Grant* is an important designation. Parent Grants are a vehicle the agency uses to allow unsolicited grants on investigator-developed topics to be submitted and to clearly designate this kind of submission from NIH-targeted topic grant solicitations. As shown in Exhibit 2.3, the information in the Parent Grant is similar to that in a focused FOA/NOFO (aka PA, NOSI) and is divided into eight sections. Most of the information you will want to convey to your colleagues in an email discussing a Parent Grant can be found in the parts of the PA titled Part 1 and Part 2; Sections 1, 2, and 3. Many funding agencies allow both investigator-developed submissions as well as submissions on a requested topic. Any submission has to meet the interests of the funding source.

GENERAL DESCRIPTION OF A FUNDING OPPORTUNITY

These three sections contain most of the information you will need to set up a meeting with your colleagues:

- Section I: Funding Opportunity Description
- Section II: Award Information
- Section III: Eligibility Information

Applications to NIH are submitted on designated forms; the grant announcement indicates which form to use. The PA chosen for the example in this chapter is PA 20-185. The 20 is the year (2020) the announcement appears, and the 185 is the specific number for that PA. If searching for announcements

EXHIBIT 2.3. Sections of a Parent Funding Announcement

Part 1. Overview Information
Key Dates

Part 2. Full Text of Announcement
Section I. Funding Opportunity Description
Section II. Award Information
Section III. Eligibility Information
Section IV. Application and Submission Information
Section V. Application Review Information
Section VI. Award Administration Information
Section VII. Agency Contacts
Section VIII. Other Information

that are recent, look for PA numbers that are close to the current year (e.g., PA 22-xxx). The NIH R01 PA chosen as an example has the activity code R01 to indicate the type of grant the announcement applies to. The R01 is the most common large grant applied for by investigators. The R01 PA has multiple options related to identifying your project as (a) not a clinical trial, (b) a basic experimental study with humans required (BESH; discussed in Chapter 3), and (c) a clinical trial. None of these approaches are new to research grants per se, but the specific designation of not a clinical trial, a BESH study, and a clinical trial is now part of the choice process, and the designations result in more emphasis on some aspects of the grant application.

The choice of a clinical trial will emphasize biographical sketches, data analysis, data management plans and data sharing, the inclusion or exclusion of human subjects, and the details of human subjects procedures combined with the methods. A detailed discussion of the clinical trial choice is provided in Chapter 3.

There are beginning and end dates for a Parent Grant announcement, but these dates do not have the import they do for research-specific FOAs/NOFOs because it is highly likely the Parent Grant opportunity will be renewed. In regard to specific interest FOAs/NOFOs, the date the opportunity opened and the date it closes indicate two things: (a) how many opportunities the applicants/investigators have to submit and (b) how long the specific announcement has been available. This is not an issue with Parent Grants, which are typically renewed.

Funding agencies provide information on how many grants they have funded on a topic through an FOA/NOFO or a PA. For NIH and federal agencies listed on grants.gov agencies this information can be accessed through the following website: https://reporter.nih.gov. This website requires some self- or other training to navigate but will reveal grants funded on a topic of interest, to include years of funding, the PA from which a topic was funded, the institution funded, and the institute official who is managing the grant for funding source (System 3). Using this source, applicants/investigators and their grant support workforce could check funded grants in their own institution (System 2) to see who is working on topics of interest to them.

For investigators new to the grant process, communication with colleagues about funding opportunities can be a learning process. The email exchange in Exhibit 2.4 is a fictitious example of how investigators can use the information from a potential funding source to quickly share the essential information provided in an announcement. In this example, an NIH NOSI—an NIH research announcement targeted toward a specific area of interest—is being referenced to provide a coherent team discussion about the potential for a

EXHIBIT 2.4. Communication Example for Discussing a Notice of Special Interest

Email 1: Initial letter from Nancy to the group

Dear John and Emily,

I found this call for research funding on the National Institutes of Health website (URL here and announcement number here), and it seems to fit with our emerging interests in the long-term nervous system effects of corona-type viruses. I spoke with Phil, who is just returning from sabbatical at UTSA Health Sciences Center, and he agrees it would be a good outlet for us.

For an R01, the limits on funding per year are capped at an upper limit of $400,000 per year direct costs. It appears that several NIH agencies have signed on, to include the National Institute on Aging (NIA), Neurological Disorders and Stroke (NINDS), Child Health and Human Development (NICHD), Allergy and Infectious Diseases (NIAID), Minority Health and Health Disparities (NIMHD). I think NIA and NINDS would be most interested in our perspective, but we do have a large sample of health data from racial-ethnic minority subjects coming in from our satellite clinics from across the state.

The announcement came out yesterday and the last date for initial submission is May 25th, which gives us roughly 7 weeks to prepare the proposal. There is a letter of intent (LOI) due April 15. The announcement extends out submission dates for 3 years, so if we are not prepared for the May 25th date, we could wait until the September date, with earliest funding in January as opposed to October. The extended date allows for multiple resubmissions.

The format appears to be a standard NIH proposal using the recent SF-424 forms. Clinical trials cannot be supported with this application, but we are using human subjects so we should be sure to look at the new forms to see how human subjects procedures are reported. The announcement focus is on sorting out potential relationships between areas of the brain involved relative to behavioral measures, symptoms of the original disease onset and offset, and the potential for identifying what type of transmission may have occurred within the nervous system. There are no limits for biological measures, although there are suggestions for use of measures that would match some existing databases. There are general goals for aims noted in the announcement, but the specific aims are up to us.

There are no limited submissions, so we do not have to prepare for an internal competition prior to submission, and our institution meets the criterion for a submitting institution. If we are going to collect data from satellite clinics across the state, we should contact the institutional review board (IRB) to see if a single coordinating IRB is our best bet and how we should proceed with IRB to set that up. I see that my current Program Officer, Gude Money, is a contact at NIA program for this FOA—so I will contact him. Can we set up a meeting no later than next Tuesday afternoon? I'll have my new postdoc, Wicked Smart, send around a Doodle for next Monday and Tuesday. I'll ask Wicked to break out the essentials for the proposal and send them around by Friday so you can look at them over the weekend.

Best wishes,

Nancy (and Phil)

EXHIBIT 2.4. Communication Example for Discussing a Notice of Special Interest (*Continued*)

Email 2: Letter of response to the group from Nancy 2 days later

Dear John and Emily,

I talked with Gude Money at NIA, and he liked our preliminary ideas and said they would be a good fit for the LOI, but we are not strictly bound by the LOI for the final submission as long as we stay within the general idea and aims. I mentioned our paper coming out in August detailing our new blood test related to predicting conversion from mild cognitive impairment to dementia status, and Gude indicated that new methods would be accepted if they were well validated and could be shown to have strong rigor with respect to reproducibility and authenticity of biological materials. He suggested staying within the more well-known neuropsychological assessments, particularly those with good and updated norms. He mentioned the NIH Toolbox at least three times in our conversation.

I will interpret that mention as a strong wish at the level of program, although for review our group doesn't have a track record with the Toolbox. I suggest we bring in Lester Metric, who is in his second year with an NICHD grant using the Toolbox. He would give us some insight and credibility with the methods and could open the doorway for some more life span applications in the future.

Gude also mentioned that the NOSI came out quickly and there was no specific contact listed for scientific review, but these applications would be reviewed at CSR as a PAR, and to check the initial NOSI for updated announcements. He knows the SRO is going to be Beon Time and that we should contact him if we have specific questions about the review format or how we should handle any appended or updated materials. He also mentioned that notices for potential R21s that might be useful for more innovative measures might be coming out from some institutes.

See you next week at the Center for Translational and Clinical Science conference room in the Precision Medicine Center Building.

Best wishes,

Nancy,

P.S. Phil is still unpacking but will be on top of things by next week.

Note. The Center for Scientific Review (CSR) and the different roles of the Program Officer (PO) and the Scientific Review Officer (SRO) are discussed in detail in Chapters 7 and 8 and are listed in the National Institutes of Health (NIH) glossary.

grant submission. The NOSI will indicate what institutes and centers at NIH are participating in the opportunity. Note how the initial email communication references each of the 10 key elements of an announcement shown in Exhibit 2.4.

These email exchanges, between experienced investigators, discuss all the points mentioned in Exhibit 2.2: (a) required focus; (b) who is funding; (c) funding limits for a proposal (to include grant mechanisms); (d) the time elements of the submission, review, and funding cycle or future cycles;

(e) the forms to be used for submission; (f) the clinical trial options; (g) any limitations on submitting; and (h) who can be contacted at the funding source about the proposal.

The fictitious exchange also shows how the announcement of an FOA/NOFO sets up the details for initial communication between investigators. For example, the first email is asking colleagues if they are interested in quickly participating in setting up meetings. The need to get moving to the next step, or to promote within-research group discussion, is emphasized, along with exploring potential use of existing resources. For example, the second email points to the potential need to access resources outside the initial group that the investigators are more familiar with (e.g., NIH Toolbox; https://www.nihtoolbox.org/). There is an illustration of navigating the funding agency via contact with the Program Officer (PO) to discuss directions that will and will not be taken (no clinical trial, no innovative measures unless well validated), recognition of pending human subjects issues, and the setup of the context for review.

The emails indicate that the choice was no clinical trial. In general, though, when communicating with your colleagues you will want to be specific about the choices of no clinical trial, clinical trial, or BESH study. Some discussion may be required to decide whether you have a clinical trial, which is why there is a decision algorithm offered online (https://grants.nih.gov/policy/clinical-trials/definition.htm).

If writing a letter to an NSF or DoD PO, ask about deadlines for a proposal. Some NSF grants have an open submission policy without a stated deadline, but the review has to be organized in some fashion. Ask the PO how the open-submission policy will affect the review process.

KEY TAKEAWAYS FROM CHAPTER 2

- The FOA/NOFO provides the key elements needed to begin developing a grant proposal.

- Applicants/investigators new to the grant writing process will find the nomenclature a challenge and will need to consult funding agency glossaries and develop their own glossaries for the most frequently used terminology.

- Different funding agencies may have their own nomenclature, but the general processes of writing a grant will overlap between funding sources.

- Deciding whether an FOA/NOFO supports a clinical trial has become an important goal. Determining whether your project meets the requirements of a clinical trial has become an important component of carefully reading an FOA/NOFO.

- ESI and NI statuses are recognized by NIH.

- The management skills of applicants/investigators are immediately engaged when planning the grant proposal, which is when the competition for funding begins.

- Part of learning how to write grant proposals is learning how, and with whom, to communicate.

3 HOW GRANT POLICY INFLUENCES STUDY DESIGN

Education is not only about acquiring and remembering information; it is also about shifting our attention. The same can be said of grant literacy. Part of becoming grant literate is paying attention to the policies that serve as a guide to grant writing, grant review, grant funding, and grant management. In Chapter 2, the focus was on reading the Funding Opportunity Announcement/ Notice of Funding Opportunity (FOA/NOFO) to know how to apply now. In this chapter, the focus is on the FOA/NOFO to find out what policy might have changed from the last submission or what changes are in the planning stage.

The first formal introduction most applicants have to funding agency policy is a an FOA/NOFO, Program Announcement (PA), or Research Funding Announcement (RFA). In Chapter 2, I explained that grant opportunities from different funding sources can have different formats and vocabulary. For all funding sources, the announcement format is both directed by, and informs applicants/investigators about, policy. Sometimes it is the experienced applicants/investigators who have to be most aware of policy change because they have prior experience with older policies and will proceed

https://doi.org/10.1037/0000390-004
Get Funded: A Practical Guide to Understanding the Grant Application Process and Writing Winning Proposals in the Behavioral and Biomedical Fields, by J. W. Elias

developing an application with those older policies in mind. Administrators involved in System 2 can sometimes stick their heads in the sand hoping changes in policy will not occur because those changes will cost money to navigate or require reorganization.

This chapter describes the key policy issues that guide grant application submissions, as well as the influence of policy on how studies are designed. Using key FOA/NOFO information, you can look for policy change in the following areas:

- who can apply (and who cannot); this can change by mechanism or grant type
- how to register to apply for institutions or for special investigator status (e.g., Early Stage Investigator)
- when applications can be and must be submitted (e.g., cycles)
- what kind of an application can be submitted
- how many applications can come from an institution for a particular announcement
- limits on submission of overlapping aims for multiple submissions
- limits of a budget and domains within a budget
- number of budget years that can be applied for, including whether grants can be renewed beyond the original submission
- what budget amounts require special permission
- forms to use for an application, including budget format
- page limits related to the actual grant proposal
- materials allowed in an appendix
- whether a Letter of Intent (or a white paper, e.g., for the Department of Defense) signifying the application is coming is required, and by what date
- the review criteria
- human subjects procedures (e.g., consent) and animal welfare procedures
- institutional review board (IRB) responsibilities and due-process dates
- rules for inclusion (e.g., racial and ethnic minorities, children, age range, sex/gender) and explanations for exclusion
- evidence of rigor with respect to handling and describing materials and procedures (e.g., genetic material, biochemicals, tissue samples)
- requirements for letters of support and when they are due and, for training grants, how are they submitted and who should submit them
- whether a clinical trial is planned and whether the research meets the definition of a clinical trial
- a plan for data management, including collection, storage, and sharing of data, and budget
- due dates and natural disasters that may delay submission
- what research is permissible with federal funds

Relative to policy, the three goals are to (a) be aware of it, (b) follow it, and (c) watch for change. One of the most important components of good management and accommodation of new information from System 3 is the ability to recognize the signs of coming change in grant policies.

The National Institutes of Health (NIH) and other federal agencies refer to, and expect stakeholders in the granting process to examine, the NIH Guide for Grants and Contracts (https://grants.nih.gov/funding/about-nih-guide-to-grants-and-contracts.htm). Exhibit 3.1 is a screenshot from the following website: https://nexus.od.nih.gov/all/2021/09/02/keeping-track-of-the-latest-nih-policy-changes/.

The following are some policy areas applicants/investigators should pay close attention to with respect to budget, development to the aims of a proposal, and the support of those aims:

- human subjects recruitment by designated status, such as sex, gender, ages of subject inclusion (children, adults, older adults, racial and ethnic minorities)
- animal welfare and administrative burden
- breadth and detail of subject consent
- review of proposals by IRBs
- rules for defining and designing clinical trials
- rules for collecting and reporting data for clinical trials, limits on budgets, and years of funding
- data management security, storage, and distribution
- improvements in the reproducibility of studies
- choosing the correct forms

EXHIBIT 3.1. Screenshot From the National Institutes of Health (NIH) Website of Guidelines for Applicants/Investigators

By **NIH Staff**

Posted September 2, 2021

0 Comments

*NIH informs the research community of policy changes through notices in the **NIH Guide to Grants and Contracts** (often referred to as the NIH Guide). These policy notices supersede information in the **NIH Grants Policy Statement** and compliance with these policy updates become a term and condition of award. Each fall, our policy team incorporates these notices into an annual update of the Grants Policy Statement.*

Here are a couple of tips to remain well-informed of current policy between annual Grants Policy Statement updates:

1. ***Subscribe*** *to our Friday emails that list all the notices and opportunity postings to the NIH Guide for the week.*
2. *Periodically refer to our **Notices of NIH Policy Changes** page (especially before submitting applications or reports).*
3. *NSF policy guidelines. https://www.nsf.gov*

A list of websites where you can obtain information on these important topics is provided at the end of Chapter 1. Applicants are used to using general, or boilerplate, information to describe general facilities and resources. In the past, the boilerplate approach was often applied to Data Management Plans as well. NIH is now requiring that Data Management Plans provide information that is tailored to the specific data and methods of collection stated in the proposed research plan.

Peer review study section participants should always take some time to catch up on how changes in policy should be responded to when writing a review. As time goes on, the requirements for reviewers typically become clearer, and reviewers provide feedback that comments on detail and adherence to policy. The FOA/NOFO provides the opportunity to lay out the components of the proposal and to estimate the need for content and the time required to produce that content. One of the advantages experienced applicants/investigators have is their greater likelihood of accurately judging the time and effort required to address policy-related issues. Experienced applicants/investigators may be able to more quickly access resources needed to help with policy changes.

Unless policy is covering emergency situations, such as natural disasters or national emergencies (COVID-19), changes in policy are always announced months, and typically several review and funding cycles, ahead of the policy implementation. Advance information helps all three systems plan. Often, System 1 investigators assume that policy is someone else's issue to manage (e.g., System 2), and System 2 administrations frequently take a wait-and-see attitude before implementing major changes in administration support and function as required by the new policy. Part of this wait-and-see policy is trying to sort out who in that system will take the lead in announcing or managing policy change. Those who leave policy issues until the last minute will have many an uncomfortable moment if there is a scramble to interpret and meet funding agency policy requirements.

SIX TOPICS IN GRANTING POLICY THAT CONTINUE TO INFLUENCE STUDY DESIGN

A change in policy can have a cascading effect on the specific components of a proposal and affect how those components are developed. In particular, applicants/investigators should consider how a policy change might affect the collection of data and the research design and analysis of those data.

Changes in NIH Clinical Trial Policy and Data Management

Changes in NIH clinical trial policy usually promote changes in other policies as well. When one funding source puts forward major changes in conducting research protocols, the ripple effect can lead to, for example, changes to human subjects review and reporting as well. IRBs, as well as changes in consent and data management and data management systems, often are significantly influenced by changes in NIH clinical trial policies. System 2 research institutions now have developed or expanded specific offices and grant management groups devoted to the submission and management of clinical trials. Applicants to NIH currently need to designate whether their project is a clinical trial versus not a clinical trial and carefully check the definitions.

Increasing pressure from federal legislative bodies and research publications has focused attention on replication of results. As consequence, reproducibility and data outcome accessibility are incorporated as signs of rigor in NIH applications. Rigor, reproducibility, and data and outcome accessibility were primary motivators of the change in the definition of a clinical trial that was put forth in October 2014. This new definition was an update to the one provided in October 2000 and is an extension of the Food and Drug Administration Modernization Act of 1997 (Pub. L. 105-115). The history of evolving clinical trial policy provides a perfect example of why System 1 applicants/investigators and System 2 administrations need to follow policy development. The development and release in the year 2000 of ClinicalTrials.gov (a complete history can be found at https://clinicaltrials.gov/ct2/about-site) was heralded as providing the organization needed for funded clinical trials (taxpayer dollars) to announce a clinical trial on a topic; explain the details of a trial; indicate whether trials are still adding subjects; and organize data for translation, provide translation of outcomes, and share data with qualified researchers.

When NIH announced its new definition of a clinical trial in 2014, it defined a *clinical trial* as one in which one or more human subjects are prospectively assigned to one or more interventions (which may include a placebo or other controls) to evaluate the effects of those interventions on health-related biomedical or behavioral outcomes. At first, investigator identification of their project as a clinical trial was required based on the definition provided, but in subsequent years NIH and Congress recognized that a significant amount of data from funded research studies were not published, data were not provided or shared, and there was no report of the final outcomes of taxpayer-supported research. So, in 2016 NIH announced how clinical trial information, as of January 2017, was to be recorded and transmitted to comply with NIH policy on reporting and sharing clinical data.

Policy often leads to clarity, but sometimes it does not. One of the more controversial components of the 2017 notice was a focus on the policy of determining whether basic experimental studies with humans (BESH) should be conducted by applicants/investigators as clinical trials.

The BESH policy has been poorly received by individuals who submit applications for the funding of those studies. BESH studies that are considered clinical trials assign subjects prospectively, by identifiable groups, not randomly. They involve an experimental manipulation that could be viewed as an intervention and often collect identifiable biospecimens that could be associated with the groups and the experimental manipulations. Investigators who conduct BESH research do not want to be subjected to the clinical regulations for reporting the study and changes in the study or for providing clinical-level data sharing. For researchers used to the culture of basic experimental studies, moving to the culture of a clinical trial entails quite a culture change in the design of a study and in data management and sharing. NIH offers a website to help with the identification of a BESH study as a clinical trial: https://grants.nih.gov/policy/clinical-trials/definition.htm.

As defined on the site https://grants.nih.gov/grants/guide/notice-files/NOT-OD-15-015.html,

> a clinical trial is one in which one or more human subjects are prospectively assigned to one or more interventions (which may include placebo or other control) to evaluate the effects of those interventions on health-related biomedical or behavioral outcomes. (NIH, 2014)

Research can be considered a clinical trial by NIH if its purpose is to understand how a basic study can be used to translate the nature of an intervention and its effects. If applicants/investigators are not focusing on application of outcomes or on producing a product or understanding the effect of a product, then they should not propose their project as a clinical trial. This sounds easy enough by basic definition, but "translation" and "intervention" and "biomedical" and "behavioral outcomes" are open to interpretation when it comes to NIH agreeing with investigators on what is identified as a basic experimental study, and not a clinical trial, and on what types of basic experimental study must be reported as a clinical trial.

Applicants/investigators can propose a BESH clinical trial and submit an application as a clinical trial if a funding source (e.g., an institute within NIH) permits BESH clinical trials. Not all NIH institutes sign on to BESH-determined clinical trial FOAs/NOFOs. Always check on current BESH policy of the individual NIH Institute.

Two conciliations offered by NIH include

- a website that provides examples/case studies and a decision process for a BESH clinical trial (https://nexus.od.nih.gov/all/2019/01/08/new-resources-available-for-basic-experimental-studies-with-humans-besh-funding-opportunities/) and

- additional flexibility with registration and reporting of results. The flexibility is extended until September 24, 2024, according to NIH (2022).

Who Are the Stakeholders in Data Management? Who Owns the Data?

Historically, an unspoken rule of science was that data management, data sharing, data security, and access to data were controlled by the Principal Investigator (PI). Data were considered by investigators to be proprietary and were not shared outside of the designated grant project team. Advances in technology allowed funding agencies to change the nature of data gathering, data management, data transfer/security, and data storage and retransfer. Changes in the definition of a clinical trial also changed the rules for making funded trial data available to the funding agencies and, potentially, to additional research stakeholders who were not part of the original data gathering process.

NIH announced the formulation of a new data management policy on October 29, 2020, and enacted the policy on January 25, 2023. Data Management Plans are not new, but the new plan is more extensive, to accommodate the new data storage and data sharing policies. Applicants/investigators are encouraged to tailor their Data Management Plans to the data collection process proposed in the application as opposed to providing generic boilerplate language (see Lauer, 2022b; see also https://sharing.nih.gov/data-management-and-sharing-policy/about-data-management-and-sharing-policies/data-management-and-sharing-policy-overview).

Grants with budgets over $500,000 have been placing data into repositories for some time. Locating such studies within your own institution and asking for assistance on postgrant archiving budgets is advisable if assistance is not readily available elsewhere. Many System 2 institutions have been working with their libraries to provide information to applicants/investigators.

Investigators producing shared databases must protect the identity of human participants and at the same time provide enough data for others accessing those data to be able to reproduce the experimental outcomes of the research. If some data cannot be provided, applicants/investigators will need to provide an explanation. The federal funding agencies have given a great deal of thought to the development of this policy, and there will be specific locations on the forms used to submit applications where the information can be provided and then uploaded. Typically, no more than two pages are expected for each plan, with budget information provided in the overall budget plan.

At NIH, Program Officers will review the Data Management Plans, but given the volume of grant proposals requiring review, they will likely review them on a just-in-time basis (see the NIH glossary; https://www.nih.gov/ health-information/nih-clinical-research-trials-you/glossary-common-terms) for just the applications that are forwarded to institutes and centers council meetings. For applications about to be funded any applicant/ investigator, corrections in plans will need to be made before a Notice of Grant Award (NOGA) is received.

With some exceptions, human subjects are also stakeholders in data processes. When investigators specifically collaborate with human subjects, the investigators' transfer of data and sharing of data must be explicitly consented to by subjects. For example, in an acknowledgment of changing research practices and interests, NIH developed a policy to allow consent to be obtained for future research with data gathered from an initial study, when the data involved identifiable private information or identifiable biospecimens.

One of the most important components of the new data management policy and its website are the directions to provide a budget for data management as part of the budgeting process for archiving data after the grant period ends. The data management website (https://sharing.nih.gov/data-management-and-sharing-policy/about-data-management-and-sharing-policies/data-management-and-sharing-policy-overview) provides examples of what can be placed in the budget for data management.

The data management budget must be expended within the term of the grant and must be used before the close-out of the grant. The guidelines for writing the data management budget and the method by which to expend the budget for the future sharing of the data are not new to the grant writing process, but applicants/investigators will need to become familiar with the new guidelines.

Applicants/investigators will need to designate a repository and a repository process for entering and storing data as well as a repository process that indicates how others can access the data. The due dates for making data available are either immediately after the first article is published or at the termination of the grant funding period, whichever comes first.

The most difficult applied components of a data sharing plan are accurately estimating the cost of preparing the data for a repository and managing the timelines of depositing the data. The most difficult psychological and social components of the policy will be adaptation of Systems 1, 2, and 3 to the new culture of data sharing. Researchers other than the System 1 PIs and the System 2 supporting institutions will have access to data that they did not write or support a grant to collect.

The time frame for providing the data in a repository and the following months could overlap with the System 1 PIs writing articles and grant submissions that are based their own data. It is not clear yet how System 2 can adequately comply with System 3 data management rules.

The Common Rule and Protections for Human Subjects

It is important to note that NIH's implementation of the new clinical trial rule on how data are to be recorded occurred within the same relative period that the revised Federal Policy for the Protection of Human Subjects was announced (https://www.federalregister.gov/documents/2017/01/19/2017-01058/federal-policy-for-the-protection-of-human-subjects). Known as the *Common Rule,* this policy was adopted by 20 federal departments and agencies as an update to the 1991 Common Rule. The goal of the 2018 Common Rule was to protect human subjects (and, consequently, investigators and research institutions) and add clarity to the rules for recruitment.

Given that there were many changes from the previous Common Rule, the U.S. Department of Health and Human Services provided a useful site for understanding the changes (https://www.hhs.gov/ohrp/education-and-outreach/revised-common-rule/revised-common-rule-q-and-a/index.html#broad-consent-in-the-revised-common-rule).

For example, in an acknowledgment of changing research practices and interests, a policy was developed to allow consent to be obtained from human subjects for future research that uses data gathered from an initial study, when the data involved identifiable private information or identifiable biospecimens. The consent process requires some thought on the part of both the investigators and the subjects. For example, in the present, the investigators must explain to the original subjects the potential future use of the data, especially if there will be potential profit. If agreeable to subjects, this is a valuable tool for investigators who might want to avoid the costly re-consent process and the loss of data that might result. IRBs are good sources to check with to see whether the investigator's (or investigators') interpretation of the Common Rule matches the intent of the Common Rule.

Recruitment of Racial and Ethnic Minority Subjects for Research Projects and Trials

A policy issue that has been in force for several years but that has been receiving more interest at the System 1 and System 3 levels is the recruitment of racial and ethnic minorities into research studies and of course clinical trials. Minorities are more difficult to recruit to research in studies because of their comparatively fewer numbers in the U.S. population, cultural knowledge of poor treatment in past studies, and because potential minority subjects are prized as recruits for research studies. Many of the changes in human subjects policies related to recruitment and consent by System 3 policies were motivated by the need to include racial and ethnic minorities in clinical studies and by the need to provide an environment of trust for participation in federally funded research. NIH has managed feedback on recruitment issues through a study section review process whereby a code can be given by study sections indicating that a change in subject recruitment is needed prior to funding a study. The recruitment issues are resolved at a System 3 funding agency level, not at a System 2 supporting institution level. Sometimes recruitment issues can be resolved after receiving funding, but it is much easier to address human subjects recruitment at the time of submission (https://www.niaid.nih.gov/research/grants-bar-awards-human-subjects).

Policies can vary depending on the granting agency and the purpose of the funding, but most often individuals grouped into the following categories can be designated as racial and ethnic minorities. In general, these categories include Black American, Native American (including American Indian, Eskimo, Aleut, and Native Hawaiian), Asian-Indian American (including a person whose origins are from India, Pakistan, or Bangladesh), Asian-Pacific American (including a person whose origins are from Japan, China, the Philippines, Vietnam, Korea, Samoa, Guam, the U.S. Trust Territories of the Pacific, the Northern Marianas, Laos, Cambodia, or Taiwan). For some purposes, individuals who are disabled or disadvantaged financially can be considered to have minority status. For research grants, the FOA/NOFO will determine minority status. Minority eligibility status for submission of a grant may not be the same as eligibility for inclusion as a subject in a research study. An emerging issue is how many subjects of each minority status are considered appropriate for a research study. At present, FOAs/NOFOs provide no guidance on how to view, for example, a sample that comprises 40% Caucasian, 35% Hispanic, 20% Black, and 5% other racial and ethnic minorities. One study section member might consider this ratio unacceptable and provide an "unacceptable" rating, and another might find this ratio acceptable. Applicants can seek help with these issues from the NIH Inclusion Policy Officer located in the Office of Extramural Research (OER) via inclusion@od.nih.gov.

Investigator Policy

Another of the more important policy changes for NIH has been the institution of Early Stage Investigator (ESI) status. NIH provides a useful URL to quickly explain the dimensions and benefit of ESI status: https://grants.nih.gov/grants/esi-status.pdf (see Chapters 2 and 7). For applicants/investigators who meet this status (within 10 years of one's highest degree of training) it is important to register and provide validation. If in doubt after reading the criteria, check with a program official. ESI status is typically conferred by NIH policy to an individual who has not received a major NIH award (e.g., R01 or equivalent) within 10 years of receiving one's final degree, including the expected degree of postgraduate clinical training. NIH provides a list of awards that will not disqualify ESI status, including training awards and career awards (see https://grants.nih.gov/policy/early-investigators/list-smaller-grants.htm). ESI status permits an advantage in the review process (see Chapter 7) and in the funding process (see Chapter 8).

New Investigator (NI) status is conferred on individuals who have not received a major NIH award (e.g., R01 or equivalent) within 10 years of the highest conferred degree and are beyond the 10-year period. Depending on the individual policies of NIH institutes, NIs may receive some benefit in funding, but there is no particular benefit provided relative to how applications are reviewed. Applicants/investigators may have extenuating circumstances that will allow an NIH institute to extend the time of ESI status. The important thing is to follow policy guidelines. On occasion, the terms *NI* and *ESI* are used to refer to the same status, so applicants/investigators should be sure that discussions of NI status and ESI status are specific to a period on either side of the 10-year post-highest-degree point of funding time frame.

Age Inclusion Policy

As further evidence of how policy influences the granting process at NIH, in a December 2, 2020, post on Inside NIA: A Blog for Researchers, Rene Etcheberrigaray noted that the increased focus on lifespan and older age as a subject-inclusion variable resulted in changes to study sections. Several study sections added expertise on research aging, and others began to focus more toward review of lifespan or older age involvement.

The NIH (2017) Policy and Guidelines on the Inclusion of Individuals Across the Lifespan as Participants in Research Involving Human Subjects explains the inclusion policy and the rationale for the inclusion policy this way:

> Section 2038 of the 21st Century Cures Act, enacted December 13, 2016, enacts new provisions requiring NIH to address the consideration of age as an inclusion variable in research involving human subjects, to identify criteria for justification for any age-related exclusions in NIH research, and to provide data on the age of participants in clinical research studies.

KEY TAKEAWAYS FROM CHAPTER 3

- In Chapter 2, the FOA/NOFO was discussed as the source of guidance for applicants\investigators who want to construct a research project and manage a submission within a granting cycle. In this chapter, the policy issues that guide submission and design of studies are revealed.

- Policy continues to evolve and influences both System 1 and System 2 functions through its influence on (a) subject safety for animals and humans; (b) human subject recruitment and consent; (c) rules for data management and sharing; (d) design and definition of clinical trials; (e) rules for what happens when a submission policy is violated; and (f) inclusion by age, sex/gender, and racial and ethnic minority/disparity status.

- Policy issues and changes are made with public notice and the opportunity to comment.

- Ignoring policy issues is a common error for both System 1 and System 2 applicants. Ignoring policies can result in applications being returned without review; after a review bars funding, whereby funding cannot be released until the policy issue is resolved (see Chapter 8); and programs being halted for Systems 1 and 2 until investigators and institutions conform to policy.

- Policy issues evolve to meet System 3 issues such as low funding for NIs, inequities in study participation, and proposed policy changes in reviews to manage inequities in funding (see Chapters 8 and 12).

PART **II** WRITING THE APPLICATION

4

GENERAL GRANT WRITING TIPS

Dos and Don'ts

Some writers rarely make errors in spelling, punctuation, syntax, or choice of words and can write without rewriting and rewriting to get it right. It is amazing. Most of us do not have that kind of skill, especially when writing under time pressure and sharing documents that are being passed around and returned with new ideas. Cutting and pasting text is an efficient use of word processing programs but also a potential source of major embarrassment.

Applicants/investigators should be trying to make a good impression on reviewers, and when reviewers find errors in spelling and grammar they may think the application was rushed or the applicants/investigators are careless and sloppy. Misuse of tense, misuse of apostrophes (possessives, plurals), and a lack of proper modification of nouns by articles ("a," "an," "the") can reduce the intellectual value of the writing and negatively influence the overall impact score. When English is an applicant's/investigator's second language, these kinds of errors are more likely. The solution for this is to seek the services of an English-speaking editor or use the editing component of a word processing program. These programs are very good at finding errors and offering suggestions for correction.

https://doi.org/10.1037/0000390-005
Get Funded: A Practical Guide to Understanding the Grant Application Process and Writing Winning Proposals in the Behavioral and Biomedical Fields, by J. W. Elias

Formal writing requires editing so that we present ourselves and our ideas with clarity. The social connection between peer reviewers and applicants/investigators is not all due to content. There are issues not related to content that can affect the clarity of an application and thus the social connection. The sections that follow address the more common errors observed in grant proposals that proofreading can quickly discern. It is easier for the writer of a document to detect errors if they can put aside a document for a few hours, or days, if that luxury exists.

REDUCE READER WORKING MEMORY

Because of the time-restricted context of grant reviewing, reviewers will gain a better understanding of a proposal if they can efficiently read the application. Writers who use long sentences or phrasing that taxes a reader's working memory can sap the reader's energy and make a document dense and difficult to understand. When applicants/investigators (writers) attempt to explain their goals and methods to others, their sentences may become long because they perceive the ideas and content of the sentence as highly related, like beads on a string. For the reader, though, who is just becoming familiar with the content of a sentence within the context of a paragraph, encountering a string of ideas may result in cognitive overload. The more related the content is for the writer, the more likely it is that they will attempt densely written sentences that express a string of ideas. Using a proofreader who is less familiar with the topic, and who is skilled in spotting dense writing, will help with this issue.

There is no standard length for a sentence, but obviously sentences approaching 40 to 50 words in a technical document could be too long, yet too many short sentences can make writing seem choppy. Writers who use a mixture of shorter and longer sentences can improve the flow of the writing. A mixture of both active (focused on actors) and passive language (focused on actions and themes) can make writing less mechanical for the reader (Ferreira, 2021). When proofreaders comment on the density of a document, the authors should examine the sentences for the number of ideas expressed and the clarity of the initial context.

Components of speech, such as "these," "those," "this," or "that," are *demonstratives* (also called *determinants*) and refer to people, places, and things. "They" (indefinite referent) can modify a noun that is present or implied in a sentence. Referent words can be useful space savers and function as shortcuts in speech and writing, and instances of "they" (indefinite) are ubiquitous in the English language to the point where "they" (indefinite) are hardly given a thought by a writer or speaker. But concise writing does not always mean precise writing, and these components of speech are problematic when the referent of the determinant is too indefinite for the context.

Indefinite references often require the reader to stop and reread sentences to know what "these," "those," "this," or "that," refers to, or who is acting. Too long a reading distance between the referent and the indefinite referent increases the demand on a reader's working memory. When there is the potential for multiple referents to match to the indefinite referent, the reader must then stop reading and check for the context of the referent. Proposal writers who are familiar with the context of actors and actions in a document may not notice when a referent is too indefinite. Proofreaders are a solution to this problem as well.

SAVING SPACE AND GAINING CLARITY

Writing guides often caution writers to edit their writing for active versus passive writing. The *active* style typically introduces the actor in a sentence early, which in many sentences reduces the number of words and thus the reading workload of a sentence. This type of edit by applicants/investigators is particularly useful for limited-page sections of an application, such as the Aims and Goals sections.[1] Much of the stylized writing in academic works uses a *passive* style that is familiar to writers and reviewers who use the familiar context of the writing to infer who is initiating action in a sentence.

The active versus passive style can sound less academic, but the active style typically uses fewer words (which are precious in a grant proposal) and provides an opportunity to insert vigor into the writing. Consider the following examples:

> Research staff will recruit subjects from the university setting and the three surrounding counties.

> Subject recruitment by research staff will take place within the university setting and the three surrounding counties.

---

[1] The best source of information regarding the various sections of a grant proposal is the Funding Opportunity Announcement/Notice of Funding Opportunity (FOA/NOFO; aka Research Funding Announcement [RFA] and Program Announcement with Special Review Criteria and/or Specific Receipt Dates [PAR]). The FOA/NOFO will refer to the appropriate forms for submission. The forms will contain the sections of a proposal, but the FOA/NOFO or a follow-up notice (NOT) to the FOA/NOFO may change how a component is addressed. The information in the FOA/NOFO or an update by an NOT overrides any general rules for writing a section.

Subjects will be recruited by research staff from the university setting and the three surrounding counties.

The first sentence uses the active style, where the "who" (research staff are action takers) precedes the "what" (subject recruitment and research staff action to be taken) and the context in which the actors act and the action takes place. The second and the third sentences focus the reader's initial attention on the "what" and then the "who," followed by the context in which the action takes place. Part of choosing whether to use an active versus a passive approach is deciding how to focus the reader's attention with the initial words of a sentence.

Writers can make sentences weighty and ponderous if they turn verbs and adjectives into nouns. Gail Radley (2019) described the verb-and-adjective-to-noun process as producing *cloggers*. This issue is also referred to as *nominalization*. A common nominalization practice in scientific writing is to add an "-ion" to an adjective or a verb, such as making *nominal*, the adverb, into *nominalization*, the noun.

Compare Sentence A with Sentence B in the following two examples:

Example 1

(A) Participants showed consider**ation** of the options and then made deci**sions.**
     versus
(B) Participants consi**dered** the options, and then they deci**ded.**

Example 2

(A) We have expect**ation** of participa**tion** by 75% of the existing patient pool.
     versus
(B) We **expect** 75% of the existing pool to partici**pate.**

In addition to *-ion*, Radley (2019) indicated that nominalization endings include "-ance," "-ence," "-ery," "-ment," "-ness," and "-son"—which suggests that we are all heavy purveyors of the nominalizing practice.

Radley (2019) offered this illuminating example of how sentences can be shortened when nominalization is detected:

The description of the horse's jump provided a surprise to the owner and a show of the skill of the trainer. (21 words)

Our nominalizations here are "description," "jump," "surprise," "show," and "skill." Notice the abundance of preposition phrases: "of the horse's jump," "to the owner," "of the skill," "of the trainer." They make me feel like I'm galloping along on that horse!

Here's a (mostly) de-nominalized version with no prepositional phrases:

The trainer described how the horse jumped, surprising the owner and showing the trainer's skill. (15 words)

My goal in providing these examples is not to encourage grant applicants/ investigators to search for and purge their documents of nominalization. The point is to show how writing can gain clarity and energy when writers attend to word choice. Chapter 5 focuses on writing as it pertains to the development of the aims of a proposal. The Aims sections in grant proposals are typically brief; sometimes only one page is allowed for an explanation of the plans for a project with multiple aims. Applicants/investigators are often desperate to eliminate just one or two words per paragraph while still retaining meaning. Reducing nominalization and using the active voice are two very good ways to create the space you will need.

Above all, applicants do not want to receive feedback that the writing is dense. The "d-word" means there was likely not a good social connection between the writer and reader and the reader and the content.

BEST EDITING PRACTICES

The best editing by applicants/investigators often occurs when there is time before submission to put the document aside for a few days (even a few hours can help). A put-aside period promotes the consolidation of ideas into more concise communication. Proofreaders can be tough on the ego, but no more so than receiving study section peer review feedback that the document contained writing numerous errors.

If multiple writers are trading and exchanging documents, be careful to save earlier versions of a document in which components still exist that were later excised from the document. That might be the best way to locate missing material that in retrospect was not that bad or is now needed. Be sure to date versions of documents and update the subject line on emails to reflect what version of a document is being shared.

The use of programs like Microsoft Word's Track Changes can be useful, but keeping all the changes and continuing to exchange documents make reading and editing too difficult. The principal writer or writers should take charge and communicate "I am now accepting these track changes."

"I will have time to write on the plane." This is a familiar statement made by rushed academics or rushed reviewers on the way to a study section meeting. The reason why the writing and the editing that are done at 30,000 to

40,000 feet in the air can seem so brilliant at the time may be due to a lack of sleep in the night or days before a flight and reduced oxygen during the flight. Conduct a postflight ground-level brilliance check before passing on the new written or edited information.

All stick and no carrot as an editor can discourage investigators regardless of their level of experience. A mock peer review session is sometimes organized to provide feedback to investigators or to a class of promising investigators being trained in how to write a proposal. Actual peer review does not involve real-time feedback to investigators, and the study section Scientific Review Officer of a peer review meeting and the chair of a peer review meeting typically control the piling on of multiple negative comments. Mock panel review can become more of a showcase of reviewer skills than a motivation session for inexperienced grant writers or new investigators. A mock review of a colleague's already-funded grant proposal can produce unintended consequences, such as convincing the grant writer they shouldn't pursue grant writing.

There is no need to sacrifice the "very good" (and "on time") to the "perfect," but there are style resources that can be used to help with writing and provide a polished proposal for review.

KEY TAKEAWAYS FROM CHAPTER 4

- Be aware that reviewers will be conscious of errors in spelling, grammar, and punctuation, and these kinds of errors may affect scoring in a negative way.

- Errors in writing may not be totally avoidable, but there should be an effort to minimize them; otherwise, reviewers may think the applicants/investigators think writing errors are acceptable.

- Avoid dense writing and sentences that increase the reviewers' working memory load.

- Writers are often too familiar with the content to see the errors. The "let-it-sit for awhile" technique is useful for self-editing, as is the use of proofreaders.

- The best motivation for applicants/investigators to use writing techniques that focus on active versus passive writing and avoid clogging by avoiding nominalization is to understand that these techniques can reduce the number of words in sentences that must be squeezed to fit page limitations.

5

HOW TO STRUCTURE THE AIMS OF AN APPLICATION TO MEET THE REVIEWER'S EXPECTATION

When an applicant/investigator develops ideas there must be, at some point, a conversation or agreement as to who "owns" the idea. Part of the granting process is that ideas and any eventual data are, at some point in time, proprietary, in particular when an invention or a product is developed from research (search the web for commentary on the Bayh–Dole Act, Pub. L. 96-517). Changes in the federal data management policies make the concept of *proprietary* a time-related notion. An aspect of developing an idea that is not often considered is who owns the idea. Ideas and data may be considered by applicants/investigators to be proprietary, but who the proprietor is, or who it should be, is not always clear. Undergraduate, graduate, and postdoctoral students most likely have a mentor who can provide guidance as students move toward independent status. When applicants/investigators ask for feedback or help when developing an application for funding, people who have contributed to the idea may feel that their contribution merits inclusion.

When applicants/investigators are working in the laboratory of an established or senior investigator, or are receiving support from a grant to these individuals, the supporting laboratory would be correct in assuming benefit

https://doi.org/10.1037/0000390-006
Get Funded: A Practical Guide to Understanding the Grant Application Process and Writing Winning Proposals in the Behavioral and Biomedical Fields, by J. W. Elias

from the support offered to students or other investigators working in the laboratory. Allocation of credit for ideas is a critical issue to sort out if you have student status, postdoctoral status, or staff scientists support and do not have status to submit as a Principal or Co-Principal Investigator. Regardless of where you are relative to independent investigator status, once you have made personal contact with colleagues and start asking for help you must evaluate how this contact will be perceived by those colleagues if you pursue an idea on your own.

GENERAL READING AND WRITING EXPECTATIONS FOR ANY GRANT APPLICATION

The fundamental audience for a grant application is the peer review study section, program officials, and upper management officials (Chapter 7 discusses the review process in detail). The best applications are well organized and meet the requirements of each funding source in a concise and precise fashion.

Regardless of the funding source, the goal of writing is to communicate and make a connection with the reader. The idea of a connection with readers and writers was discussed in a production called *The Booksellers* (Young, 2019). This program focuses on antiquarian booksellers and their clients. In a conversation about why a market exists for the booksellers and their acquired books, there is a discussion of how readers choose books. The storytellers in the titular Booksellers program propose that when a social connection is made between a book and its reader, it can be reasoned that the book chooses the reader and reads the reader, while the reader reads the book. In other words, the content should be enjoyable to the reader. (See Landrum, 2021, on how to tell a good story with scientific information.)

Research applicants would do well to try to make a social connection to the reader/reviewer and the funding source when explaining the goals and aims of an application. Part of establishing a connection with a funding source is being on target with the goals of a funding opportunity.

FOCUSING ON THE AIMS OF AN APPLICATION IN THE INITIAL EXPECTATIONS SECTION

The short Aims section required by most funding agencies is the initial platform for the concise presentation of specific and linked goals designed to advance knowledge within specified domains of science. The aims, not the applicants/investigators, are the main characters in your application. As main characters, the aims are the foundation for what must be supported by the remainder of the application. For some applicants/investigators the Aims section is an easy page to write, but it would not be unusual for them to find this section problematic or experience that some aims are difficult to support by the remainder of the application.

Writers, when first crafting a project's aims, tend to bring to consciousness multiple ideas and thoughts associated with all the components of an application. It is not unusual for applicants/investigators at all levels of experience to require several drafts of aims to pare ideas down and include just the essentials of the research proposed. Chapter 8 focuses on reading a Summary Statement and includes a discussion of how reviewers comment on aims.

Applicants/investigators can initiate the process of writing aims by stating in one sentence what is broadly expected to change in a field of science when the research is completed. This initial step establishes the broad framework within which the research goals will contribute to progress. If the specific aims stray from collectively supporting the broad statement of change either by being too narrow, or too expansive, then the applicants should rethink the fit of the aims with the overall research goal.

I recommend that investigators develop a Quad Chart (Figure 5.1) as a means by which to project how the aims can be supported by the methods. The Quad Chart is not part of the submission, but placing ideas into such a format helps organize the basic components of an Aims page. The chart can be created in PowerPoint by selecting the "Shapes" option. A large square or rectangle can be divided with lines into four quadrants. The Quad Chart shown in Figure 5.1 could have sections expanded to allow for more content.

FIGURE 5.1. Aims Quad Chart

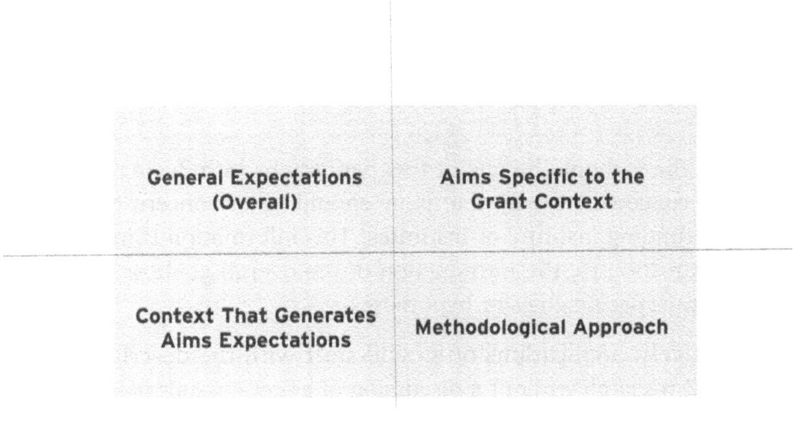

General Expectations (Overall)	Aims Specific to the Grant Context
Context That Generates Aims Expectations	Methodological Approach

In the upper left quadrant, state the general context that is guiding the need for research developed in the Aims section. Explain the need that the proposed research will meet. In a second quadrant, provide the more specific supporting context for the research proposed. For example, provide, in compact form, recent findings that are guiding the application. In a third quadrant, write the specific goals in terms of aims. In the remaining (fourth) quadrant, identify what will be needed in the approach (methods) to support the aims. This fourth quadrant could include a bulleted list of research supports required to pursue the aims, for example, design and analysis support; clinical trial or subject recruitment support; pilot data; and/or access to materials, space, specialized personnel; diagnostic materials, procedures, or devices. Follow this with one sentence for how each of these needs will be met. Adjust the quadrant size of the Quad Chart to meet your writing needs. Use the quadrant organization to consolidate and organize the components of the aims. The contents of the Quad Chart should guide the writing of the Aims section.

Initially, the components of the Quad Chart may be broad and a bit fuzzy, but as the application develops the abbreviated contents of each quadrant should be clearer, more cohesive, and easily descriptive of the overall project development. If the abbreviated components of the Quad Chart need more focus and support as you read and explain them, that indicates the application should be more focused. The Quad Chart can be used to develop the Abstract section of an application.

SUGGESTED ORGANIZATIONAL SCHEME FOR AN AIMS SECTION

Applicants new to the process of developing the aims of an application often wonder how the Aims section should be organized. The following list suggests an organization that would be expected by the reviewers of an application:

- Initially, there should be one or two paragraphs introducing the general issue of concern and why that issue should be of concern to reviewers and the funding institute or institutes. This information is in support of, and a transition to, the introduction of the overall goals or hypotheses, but not the specific goals or hypotheses.

- Alternatively, applications often will start with the description of the proposed research without a discussion of general issues promoting that research. In this approach, motivation is implicit, for example, "There is much interest in following patients diagnosed with COVID-19 who have shown initial signs of brain involvement (e.g., loss of smell or taste, to see if long-term cognitive damage has incurred)." In this case, the overall hypothesis is developed from the observation of changes in sensory input for some COVID-19 patients, but there is no theory or data available to cite to support the observation, nor is there any information on how many patients have been affected that should be presented to motivate the hypothesis. The goal of the research is important because of the newness of the topic and the motivating implications of a long-term effect on the brain when the initial symptoms are loss of smell and taste. Broad support of the overall goal, for example, the number of expected individuals to be followed and in what age groups, is then presented.

- In general, within the first few paragraphs of the Aims section the reader should be introduced to the general or specific premise for the aims and provided with some motivation as to why the general aims should be pursued.

- The premise is typically followed by transitional material, introducing the methods of approach for the specific aims and goals and offering broad support for these methods, if possible (this is typically not detailed).

- The specific hypotheses, goals, or outcomes are typically presented in the next-to-last paragraph of the Aims page.

- The last paragraph of the Aims page may be devoted to providing some additional detail about the specific aims or restating the importance of the aims in terms of their expected impact.

It is strongly recommended that the consideration of human subjects issues be part of the initial grant preparation process as the aims are being developed given that inclusion (age, sex, racial and ethnic minority status, children) and human subjects, especially in clinical trials, have become top issues of consideration for funding a grant application. In other words, be aware of how the selection and recruitment of subjects will affect the overall aims of the research. Likewise, if animals are to be used, be aware of the focus on sex differences and the growing interest in age as a biological variable of interest. (See Chapter 6 for a broader discussion of this issue.)

Applicants should take note that aims, though important, do not act. Investigators act. Therefore, Aim 1, 2, or 3 cannot show, provide, or investigate, as in "Aim 1 will show, provide, or explore. . . ." A better statement would be "We expect Aims 1, 2, and 3 to . . ."

WHAT TO LOOK FOR IN NIH-PROVIDED SAMPLES OF FUNDED GRANTS

Examining full grant application examples is useful, especially if the examples were funded. The NIH website offers several funded grants for review by the public through the NIAID web address (https://www.niaid.nih.gov/grants-contracts/sample-applications). Make certain the URL information is current and is actually vetted by the funding organization. Examples from the National Science Foundation (NSF) can be found online, but be certain that the example is clearly an NSF application, and note the year and the submission format. The Department of Defense (DoD/DOD) and other agencies can be contacted for potential examples of up-to-date funded applications—if they will provide them. Clearing funded applications for public review is a difficult task.

Readers of these examples should be mindful that the now-funded applications were submitted by System 1/System 2 applicants using the format and policy conditions that existed *at the time of submission*. The funded application samples may not include more recent forms or directions (e.g., directions for clinical trial development or database management). The funded applications provided by funding agencies do not show the reviewers' commentary, and the components shown in the examples may not all have been judged at the highest level. The overall presentation of the components in the samples of funded grants should be considered at the highest level.

Sometimes the aims of an application are the same as the aims of the Funding Opportunity Announcement/Notice of Funding Opportunity (FOA/NOFO). When FOAs/NOFOs are not Parent Grants but topic-targeted announcements, then the aims must support the goals of the FOA/NOFO. When an FOA/NOFO provides the overall goals for an application, then the objective of the applicants/investigators is to put methodological life into those general (or sometimes extremely specific) FOA/NOFO goals. Nevertheless, it is always a good idea for applicants/investigators to specifically state how the premise of the application is meeting the demands of the FOA/NOFO. Reviewers looking at multiple grant applications that are not FOA/NOFO focused need to be reminded that the specific grant application they are reviewing is focused on a specific FOA/NOFO goal.

MIND YOUR AIMS: WHAT REVIEWERS LIKE TO SEE IN AN AIMS SECTION

In addition to the expected organization of aims, the reviewers often have additional expectations for the Aims section.

- The applicants' aims should seem manageable. If the Aims section is too complex in sentence structure, or is densely written, then the reviewers may question the applicants' ability to manage the aims or manage them within the time frame of the research application. This leads to the issue of the number of aims. The concept of "manageable" for the reviewer often extends beyond the complexity of the aims to the perceived capability of the applicants/investigators, in particular the Principal Investigator, to manage and complete the project with the resources available. Be aware that complex aims must be supported by the biographical sketches and the description of available resources.

- The application should have clarity of purpose and motivation of goals, including why the application is important.

- Be sure to include what is at stake for a field of research if the research is completed or not completed and who the primary stakeholders for the findings are.

- Explain what decisions were made that led to the proposed aims and, if methods are important to the goals, the choice of methods.

- Include specific statements of the individual aims in the form of outcomes/testable goals/hypotheses and the expected outcomes.

- Explain aims that are integrated but not cascading in success (e.g., moving on to Aim 2 or 3 does not require that the hypothesis in Aim 1 be supported, and moving on to Aim 3 does require that the hypothesis in Aim 1 or Aim 2 be supported). (See Chapter 7 for a discussion of cascading aims.)

- Provide succinct descriptions of the data to be acquired, including how data will be acquired and from whom.

HOW MANY AIMS ARE REQUIRED?

The more aims there are, the more time and money are required by applicants/investigators. When reviewers expect pilot data, know that more aims require more pilot data. If the aims are complex, then the writing must become more succinct, to keep the overall document within the established page limits. The funding mechanism should support the number of aims in terms of the money and time allotted. For example, $175,000 for 2 years will not support multiple aims. Four aims, for example, would require quite a bit of pilot data and researcher time to seem feasible, and all the aims should be strong; a weak aim can detract from the overall impact and value of the submission. Two strong and well-supported aims would have a better chance of receiving a stronger review than two strong aims and one or more weaker, less easily supportable aims. Sometimes a Program Officer (PO) can provide guidance on whether two aims would be sufficient for the proposed grant mechanism (e.g., R01) that will support the research.

Applicants/investigators often propose the third aim as an "exploratory" one. Reviewers may question why an aim is described as exploratory if exploratory research is not proposed in the FOA/NOFO. If applicants/investigators think additional data will be available within the context of testing the basic hypotheses, but there is little in the literature to guide a discussion of the additional data, it would be best to describe the exploratory aim in that fashion. For applicants/investigators these new data might be best explored as an exploratory aim, but they should check the FOA/NOFO carefully for any restrictions on aims and check with the PO who is mentioned in the funding opportunity at the end of that document.

Applicants/researchers should ask themselves whether the exploratory aim is (a) a weak third aim designed to lengthen the grant and funding period or (b) an aim that is begging to be explored within the context of the first two aims. Reviewers and POs will often focus on an exploratory aim to see if it is groundbreaking enough to be considered as part of the stronger package of aims. Be aware that NIH's policy on clinical trials indicates that having just one aim meet the definition of a clinical trial can result in a project being formally identified as a clinical trial protocol.

When the FOA/NOFO provides the general or specific aims, the applicant/investigator will need to repeat those general aims in the application and then fill in the detail to show how those general aims will be pursued. When the aims are guided by an FOA/NOFO, it is wise to point out throughout the application how the FOA/NOFO is guiding the application. Reviewers who have been evaluating standard FOA/NOFO applications might become confused when the FOA/NOFO alters the rules for submission for a specific application.

The following are six common mistakes to avoid that are specific to writing the aims:

1. Applicants/investigators should be aware of, and avoid, *aims drift*. Aims drift occurs when the details of the application, usually found in the Approach section, reveal there are more aims to be assessed, or available for testing, than noted on the initial Specific Aims page. It is always a good practice to have an unbiased reader scan specifically for aims that are developed in the Approach section but do not appear in the Aims section. The goal is to avoid the following reviewer comment: "There are

three proposed aims, but the document reveals two other aims that could be examined as well."

2. Applicants/investigators should not overpromise in the Aims section and underdevelop an aim in the Approach section. This can happen when one of the aims is presented as having multiple pathways or outcomes and the detail of the methods does not fully support the multiple-outcome aim. Be aware that if the aims do not match the general aims provided within a non-Parent Announcement the application can be returned to the applicants/investigators at the referral stage, when applications are checked for compliance with policy. NSF is particularly adamant about being on target with the goals of the funding announcement. Granting agencies often ask for Letters of Intent or a white paper that describes the goals of a study before a full submission is made. This process should be taken seriously.

3. In some circumstances, applicants/investigators face a procedure that needs to be developed or improved for the proposed research to be conducted properly. Such procedures should not be part of the aims unless they are the focus of the research. If the procedure to be developed affects multiple aims, then making the procedure an aim is risky business for review and for completing the proposed research. When applicants/investigators have a procedure as an aim, then there is the potential that reviewers will think the research has not been developed to the point where it should be.

Nevertheless, when there are procedures that will need to be developed it would be better, in most cases, to place them in the Methods section and support the promise of developing the procedure in this section. When applicants/investigators focus on procedures in the Aims section, that draws the reviewers' attention to the procedural issues and not to the testable hypotheses. Applicants/investigators may want to wait to develop a procedure if doing so will result in a stronger application. If the FOA/NOFO states there is only one opportunity to apply, then submitting an application without a developed procedure might be the best choice, but the applicants/investigators must realize that the aims cannot be completed again until the review period for that application has been reviewed by a study section.

4. NIH policy dictates that applicants/investigators should not complete the same or highly similar aims more than once during a review cycle without specific, verified funding agency permission. When a Scientific Review Officer (SRO)/Scientific Review Administrator (SRA) releases a Summary Statement, applicants/investigators can see how the submitted

aims were reviewed, and a new cycle can begin. The aims can be completed after release of the summary statement as part of a different grant mechanism. The NIH referral process is clear about not using multiple mechanisms to submit the same aims (even if the number of aims is reduced) until a review cycle has ended. Applicants/investigators should reach out to the PO of any funding source if they are not sure about the overlapping aims policy relative to a specific application. Overlapping aims can be spotted during the referral process used to assign grant applications and a application may be returned without being reviewed, regardless of what the PO is willing to accept.

5. Once the aims have been formulated, applicants/investigators should not wait to begin to develop the statistical design and analysis section. Each aim must be supported by a design and an analysis. Designs can include bias or an incomplete test a hypothesis if not considered carefully. This is the point of completing the aims Quad Chart. Analytic procedures have data distribution requirements and may be limited in regard to evaluating some hypotheses. When researchers select statistical designs, sample size, power, and effect size are three elements that will be affected by the statistical design, choice of analysis, inclusion/exclusion criteria, and capability to recruit subjects or support the cost of recruiting subjects. Last-minute requests for statistical help can result in generic-sounding design and analysis sections that are less competitive compared with applications that have more detailed discussions. Worse than not being funded is being funded and then discovering that the hypotheses cannot be evaluated once data have been collected. If you were going skydiving, would you wait to have the parachute checked on the plane? Be sure to check your statistical jump procedures and statistical parachute pack on the takeoff to submission.

6. When writing grants to receive support for novel ideas one should also check with the potential funding agency to see whether the novel idea will be supported by that funding agency. Agencies often will provide information about past funding via a website. NIH, for example, has a reporter website (https://reporter.nih.gov/) where one can search for funded projects by topic, study section, institute, and year of funding. A more direct way to check for "funding novelty interest" is to contact a the funding agency's PO. A review of Notices of Special Interest (NOSIs) by category may reveal a specific interest in a topic. A sample email to a PO provided in the Appendix shows how to get to the point, with specific information.

WRITING AIMS FOR A RENEWAL GRANT

Grant outcomes are not always supportive of the initial aims and hypotheses of a funded application (see Chapter 8). Applicants/investigators may need to decide whether the new aims proposed for the renewal application are supported by the data from the previously funded application. If there are no supporting data, applicants/investigators may consider submitting a new application, not a renewal. If the data in support of the new application are from the Parent Grant, then a renewal may be the best path to follow. Some grant mechanisms are not renewable; applicants/investigators should check the original FOA/NOFO.

Reviewers and POs will expect a renewal to be developed on the basis of pilot data of publications from the previously funded grant application, in particular, if applicants/investigators are competing for a grant mechanism at the level of an NIH R01 grant. The bar for previously funded R01 investigators can be set a little higher by peer review sections, and the funding lines for previously funded applicants/investigators are set higher by funding institutes at NIH. Renewal applicants should be aware that study sections for NIH grants receive the previous summary statements from the initial funded application.

Reviewers of renewal grants evaluate the previous aims and progress as well as new aims and supportive data. The benefit of submitting as a renewal is realized only if the previous application was successful in supporting aims and hypotheses or if it can be shown that the pursuit of previous aims contributed to innovative ideas. If successful pilot data or published data from the initial grant are used, then a renewal is the way to develop the new ideas. If a new direction emerges from the previously funded application that is supported by the previous application but does not extend or support the aims of the previous application, then a new application rather than a renewal with the baggage of previous reviews may be the better strategy for developing the latest ideas. Sometimes funding agencies can provide insight into a better strategy.

Applicants/investigators seeking to renew initial grants may need a renewal to keep research going but do not have the supporting data collected, analyzed, submitted, or accepted for publication. It is especially important for a midlevel investigator to allow the data from the initially funded application to develop and be reflected upon before resubmitting. Data from grant investigations can take some time to percolate and reveal the findings or the direction for the next application. Unfortunately, the pressures to continue a stream of funding can hurry a decision to renew with new aims when it would be better if more time were taken to reflect on the previous results. If important publications were supported by the initial grant, it can be useful (but requires time) to see how others are responding to the publications and how they review the work. In Chapter 12, I suggest that a plan for a stream of funding should involve a variety of supports so that when one grant ends other supports can provide the time needed to prepare a renewal application. Years after a grant has ended and the initial results reported, an idea can come to mind as to what the data really meant, or what should have been pursued next.

MAKING A SOCIAL CONNECTION WITH THE READER

I suggested at the beginning of this chapter that applicants/investigators think of writing as making a social connection with the reader so that they enjoy reading the application. The social connection between reviewer and writer is not all due to content; there are noncontent issues that can affect the clarity of an application and, consequently, the social connection as well. The writing tips offered in Chapter 4 to help reduce errors in spelling and grammar can help one make a better connection with the reviewer.

MAKING THE CONNECTION WITH THE PO TO DISCUSS THE AIMS

Experienced and new investigators often wonder if it is best practice to cold call a PO, SRO/SRA, or other granting agency administrator regarding the aims for a new grant application. If time is of the essence, or emails have been unsuccessful in establishing communication, then perhaps a call is the best approach. In many cases, an unexpected phone call can catch an administrator off guard and lead to a requested email. Some funding program contacts prefer a phone call and will tell you so when you contact them by email. If a phone number is offered for phone contact, and not an email, then phone call it is.

An email will allow the funding contact to think through the email; answer when prepared; do research, if necessary; and answer from a variety of locations and at a variety of times. Funding contacts sometimes have their own mode of providing information that must be interpreted to gain a full understanding. Therefore, do not end an interaction with a funding institute official until the requested information is clear—even if the encounter is intimidating. When contacting an NSF or DoD PO, be sure to focus on how

EXHIBIT 5.1. Sample Email

Dear Dr. Oz:

I am a new research faculty member at the Brick Road Foundation in Toto, Kansas. I am developing a grant application that focuses on [specific science topic, e.g., memory training] and will address [science topic, e.g., working memory training] as it relates to [greater science issue; e.g., health literacy]. Specifically, I aim to improve patient understanding of (1) [Specific Aim 1; e.g., disease-specific e-health information from multiple websites], (2) [Specific Aim 2; e.g., patient actions based on memory for e-health information], and (3) [Specific Aim 3; e.g., patient confidence in health decisions after working memory training]).

I think that my research idea falls within NIA's interest as described in FOA/NOFO No. 2345222 - [title of FOA/NOFO]. I notice that this PA was issued 3 years ago, and I am wondering if NIA interest is still high in that area.

I would appreciate any feedback you can provide on NIA or broader NIH interest in my topic.

Sincerely,

Dorothy Gale, PhD

Brick Road Foundation

Toto, Kansas

Note. If there is no program announcement (PA) on your topic, inquire about interest of institute in your topic. A new PA indicates interest is still high. Explain in one or two sentences the significance and innovation of what you are doing, for example, "The project is significant because it addresses the growing presence of and reliance on web-based health information. It is innovative in the attempt to train a specific cognitive process to improve understanding and execution of health information." If there is something important about the methods, mention them briefly.

the application should be designed to meet the program goals stated in the Program Announcement. When interacting with an NSF PO, ask for specific examples of how to write in the Greater Impacts section that is required for NSF applications.

Exhibit 5.1 is a fictitious email to a PO introducing the applicant/investigator and the context of the application aims.

KEY TAKEAWAYS FROM CHAPTER 5

- The Aims section is a short one, but it must convey, in a precise and concise fashion, the greater supporting context of the research proposed, the more specific context of the research, and the specific goals. The methods may be briefly mentioned if important.

- It is worth taking some time when writing this section given that the remainder of the application must align with the aims and support the aims without wandering off track.

- The aims can reflect a researcher's or research team's de novo research interests. Funds are increasingly being provided by funding agencies to reflect the agencies' focused research goals. In this circumstance the general aims are those of the funding institute and are usually outlined in an FOA/NOFO. The specific aims of the applicant/investigator must support the goals of the FOA/NOFO.

- The aims are the characters in the story to be developed in the remainder of the application. The characters should not be hazy.

- Beware of aims drift in the subsequent sections of the application. Be certain that new aims are not being casually developed in the Approach section of the grant application.

- Aims should be interrelated, but the completion of an aim should not require the success of a previous aim or another aim.

6 PROMOTING THE AIMS IN THE SIGNIFICANCE, INNOVATION, AND APPROACH SECTIONS

Most funding sources will require several pages that support the aims in terms of why they are important and ask for evidence that supports their importance. If the aims are innovative, that could add to their importance to reviewers and funding sources. Some funding agencies, such as the National Science Foundation (NSF), will state that incremental aims are not likely to be funded.

The methodological approach used to support the aims must be discussed in a manner that shows the rigor of scientific methodology and amplifies the significance of the Aims section. Depending on the funding source, the sections of the application used to support the aims will have different names and allow differing numbers of pages, but this is how the characters of the story introduced in the Aims section are developed. The National Institutes of Health (NIH) provides specific sections and limits the pages allowed for the development of the aims, significance, innovation, and approach. The ensuing discussion of the NIH components used to support the NIH Aims section is applicable to most research funding opportunities.

https://doi.org/10.1037/0000390-007
Get Funded: A Practical Guide to Understanding the Grant Application Process and Writing Winning Proposals in the Behavioral and Biomedical Fields, by J. W. Elias

WRITING THE SIGNIFICANCE SECTION

The Significance section has evolved via System 3 NIH policy to require two elements: (a) presentation of, and support for, the premise of the study; and (b) the discussion of what is expected to change in the research field when the proposed data have been collected and analyzed. Applicant/investigator support of the premise can appear in other sections of the application as well but must be addressed in the Significance section. Given the System 3 policy emphasis on what should be in the Significance section, the importance and potential impact of a proposed research study is more likely to be rated as high if attention to a policy designed to improve rigor is observed.

If you were a lawyer and the aims were your clients, the Significance section is where support for your clients' case begins. Applicants/investigators are expected to use the Significance section to show how the literature supports the premise of the aims and can provide substance to the proposed changes in the field or to validation of existing theory and methods. Proof-of-concept publications by the applicants/investigators are very helpful as a source of vetted pilot data.

For clinical studies, applicants/investigators should address how the proposed treatment methods will improve outcomes or address issues in current clinical approaches. Will diagnosis become more accurate, will it occur earlier in the progression of a disorder, will it be done more efficiently, will it improve point-of-care treatment? Will treatments become more applicable to a broader range of subjects, or will they become more easily adhered to or more easily evaluated? Often, the best way to examine the efficacy of proposed clinical methods is to compare them with current standards of treatment, or with no treatment. Questions to help applicants/investigators evaluate clinical trials and research involving human subjects research are given in Exhibit 6.1.

For more mechanistic cause-and-effect studies, applicants/investigators can use the Significance section to validate or show improvement in methodology/ instrumentation and promote the need to explore new biological pathways or interactions within pathways. Mechanistic studies often promote the examination of variables in a combination that can solve existing quandaries. The NIH policy focus on rigor fits well with mechanistic studies when the focus on inclusion variables can provide new insights into the interaction of sex, gender, ethnic status, or age as they effect outcomes. The length of the Significance section depends on the degree of effort needed to support the premise.

When funding agencies become more aggressive in suggesting areas of needed research via focused Funding Opportunity Announcements/Notices

EXHIBIT 6.1. Human Subjects and Clinical Trials Information Form

Use the following four questions to determine the difference between a clinical study and a clinical trial:

1. Does the study involve human participants?
2. Are the participants prospectively assigned to an intervention?
3. Is the study designed to evaluate the effect of the intervention on the participants?
4. Is the effect of subjects being evaluated result in a health-related biomedical or behavioral outcome?

Note that if the answers to the four questions are yes, your study meets the NIH definition of a clinical trial, even if

- you are studying healthy participants;
- your study does not have a comparison group (e.g., placebo or control);
- your study is designed to assess the pharmacokinetics, safety, and/or maximum tolerated dose of an investigational drug;
- your study is using a behavioral intervention;
- only one aim or subaim of your study meets the definition of a clinical trial.

Studies intended solely to refine measures are not considered by NIH policy to be clinical trials. Those that involve secondary research with biological specimens or health information are not clinical trials.

For case studies, see https://grants.nih.gov/policy/clinical-trials/case-studies.htm#collapseS2_NIDDK_8

---

of Funding Opportunity (FOAs/NOFOs; e.g., women's health), the presence and the language of the announcements can be used to support the premise of the research as well as to provide direction for the aims. As further evidence of how policy influences the writing of research applications, in the January 28, 2016, edition of the NIH blog *Open Mike*, authored by Mike Lauer, Lauer announced that the focus on rigor in scientific grants would, as an official policy, require specific support for the premise of an application (see Lauer, 2016).

The goal of the System 3 policy is to have the investigator or investigators vet the evidence used to promote the proposed aims and premise identified in the Significance section. Lauer (2016) signaled that a policy change was coming that could be implemented immediately but that would be enforced as a policy beginning with January 25, 2019, applications. The NIH began to focus on the premise as a part of the overall attempt to strengthen the rigor of proposals to make outcomes more replicable. It was not clear to many experienced applicants/investigators that promotion of a premise is a new idea; the policy announcement more likely directed the peer reviewers' attention toward commenting on the premise. The System 3 focus on replicability is of parallel and primary interest to the research publication industry; a search of the published research literature will find numerous articles on this topic, including numerous perspectives on the depth of concern and how to make science more replicable (see Aliouche, 2022).

An added component of interest in the reproducibility policy announcement was a focus on biological variables, including age, sex, weight, or other biological factors that would be known to affect the outcomes of a study. Historically, sex differences were not a focus of animal studies unless specific sex-related hypotheses were being pursued. The NIH policy focus on including sex differences in animal studies has been more nuanced with respect to specific main effect or hypothesis development because there has been too little sex difference research with nonhumans to support specific hypotheses or predicted interactions. Some Research Funding Announcements (RFAs) will specifically address show sex differences in animal studies can be approached relative to developing hypotheses. As more studies are published that support sex differences in animal research, the sex difference hypotheses will become clearer, as will the strength of any interactions or main effects that those differences would be expected to produce. Clearly, one way for investigators to determine what sex difference hypotheses they should pursue is to review the literature for evidence of the strength and rigor of pertinent outcomes.

Animal researchers have had to make more of an adjustment to include sex differences as part of their research than have human subjects researchers because of the cost of adding males and females to animal studies. In addition to greater numbers of animals maintained and tested, that cost includes separate caging, separate vivarium rooms and air flow control, separate testing apparatus, knowledge of hormonal cycling and parity status, and variation in activity levels, to mention a few of the more recognized requirements for sex differences research. Researchers who use animals as subjects have for many years relied on pilot studies that included one sex and small sample size numbers to manage the cost of gathering nonfunded or low-funded preliminary data.

Preliminary/pilot studies are often underpowered, which is becoming more of an NIH policy concern because such studies provide support for the rigor of aims and the premise of a set of aims. Nevertheless, even small sample pilot studies can validate procedures and provide measurement variance estimates. The variance estimate shows evidence of how larger sample sizes could satisfy a stronger test of a hypothesis that has a specified effect size. Applicants/investigators will find that they will fare better in review if they do not promote pilot data beyond what those data can support. Many reviewers will recognize the pilot data issues and interpret the data for what they can provide.

Age inclusion is an additional complex issue to navigate, not only because of recruitment and sample size issues but also because age, like sex, is a complex variable. Age can represent time, an accumulation of experiences, or a cohort of experience, as well as expected or unexpected biological changes that represent resilience or vulnerability. Inclusion of an age range in a study requires not only more resources in the form of time and money but also attention to statistical analytic procedures that may incorporate both cross-sectional and longitudinal measurement. The federal policy of age inclusion does not specifically designate age as a group variable measured in a cross-sectional sense or age as a longitudinal variable measured across time, or with two or more repeated measurements.

With age now included as a noted group for consideration of inclusion in a study, such as inclusion of children and the sex of a subject, both investigators and reviewers must carefully consider how the literature promoting a premise treats these important variables.

Applicants/investigators who review a body of literature that reveals no, or limited, effects of biological variables important to System 3 (e.g., sex, age, minority status, disparity) should note whether biological variables have been poorly investigated (e.g., sex differences in animal research). Noting that there is limited evidence to guide current research is quite different from saying there is no evidence in the research literature. This is part of showing rigor when promoting the premise and is another example of how a change in inclusion policy could prompt investigators proposing pilot studies to determine whether their experimental design should include a check for interactions.

As noted, NIH has determined that the premise of an application, as defined in the FOA/NOFO and by referral to policy, should be defended and promoted by applicants/investigators in the Significance section. The premise should be defended and promoted by applicants/investigators throughout the proposal, but the primary process of describing the quality of the research that supports the aims of the project begins with the Significance section.

In some proposals, researchers will promote the premise by pointing out that previous research used to promote practice methods or theory is not strong. Applicants promoting the lack of quality in previous research as a rationale for new aims can put themselves in a precarious position if the reviewers (or their mentors) are likely to have provided much of the previous research. For NSF proposals, the combination Program/Review Officer may be actively publishing on the proposal topic. One way to approach this issue is to discuss how the proposed research will build on the findings and building blocks of previous research while showing that different methods may enhance the importance of the previous findings. In general, it is not good practice for applicants/investigators to promote the premise of proposed research by declaring that all that has gone before is lacking in quality.

The best overall promotion of proposed research by applicants is the use of pilot data, especially if the pilot data were collected and published by the applicants. The pilot data could come from the analysis of existing data sets that have been made available to provide access to previous research output. Investigators who use existing data sets can provide useful preapplication publications that support a premise. Use of System 3 existing data sets as a resource to apply for funding is an effective use of System 1, System 2, and System 3 resources.

If applicants are proposing the use of existing databases, or cross-walking data between existing databases, then the integrity of the database and the applicants'/investigators' experience using the databases should be promoted. Almost all existing databases can be criticized for not including some aspects of data researchers would ideally like to see. Part of researching the literature to support the use of an existing database is to find out what those criticisms are and to see whether the criticisms apply to the proposed use of that database. A mixed study involving the collection of new data can sometimes be part of the proposal when there is a missing or questionable component of an existing database.

Peer reviewers are more likely to accept investigators' use of existing databases if personnel on the research team have access to and can show experience with the database. With the advent of the new data management policy, the use of existing repositories based on previously funded grants might become more acceptable as pilot data.

WRITING THE INNOVATION SECTION

Some RFAs focus specifically on applying innovative methods to a precise problem or submitting investigator-initiated aims that are focused on the novelty of the idea. For example, the NIH Directors' Pioneer Award Program (see https://grants.nih.gov/grants/guide/rfa-files/RFA-RM-20-011.html) specifically requests that investigators depart from their usual areas of interest to address a new area of interest by essentially "turning it on its head" to bring a new perspective to the area. For the NIH Directors Pioneer Award Program, an innovative approach to an established area of research is the motivation for funding.

For Early Stage Investigators, NIH (2020) established the NIH Directors New Innovator Award Program (see https://grants.nih.gov/grants/guide/rfa-files/RFA-RM-22-019.html). The focus here is on new investigators who are "particularly innovative." In these targeted innovation-focused RFAs a substantial portion of the impact score will be determined by the judged novelty and innovation. The submission format of the New Innovator Award is essay, and the formal presentation of aims by applicants is discouraged. Preliminary data do not have to be provided by applicants in these New Innovator applications. The New Innovator Awards are also considered to be in the high-risk category because the focus on the outcomes without a focus on an established premise or pilot data.

It is not clear what "particularly innovative" refers to as opposed to "ordinarily innovative," but the assumption is that the reviewers will know it when they see it. The instructions for review are the usual ones, except for the format. There seems to be an implicit assumption in the new format for the New Innovator Award that following the basic guidelines for a regular grant submission will stifle the creative process.

For most RFAs where innovation is not the specific goal, the Innovation section is briefer in presentation than the other sections. In many applications the idea of innovation is discussed, but there typically is no effort to methodologically validate innovation as an aim. Applicant/investigator aims can be developed or supported by (a) proposing an innovative perspective; (b) applying innovative methods; (c) arguing that the outcomes, as proposed, could lead to an innovative understanding of a particular topic; (d) explaining that an innovative application of outcomes will be a result; or (e) arguing that outcomes will foster innovative perspectives that have not been previously considered. Investigator innovation carries weight only if the innovation can be introduced within a dialogue that persuades the reviewers how innovation will contribute to impact. Applicants'/investigators' key issues to promote innovation are what is new, novel, or different in some aspect of the approach to the aims, design, or expected outcomes of a study.

If innovation is not the focus of a set of aims but can and will be a component of the aims, then extracting what is new, novel, or different will supply the emphasis required in the Innovation section. In general, applicants/investigators should not oversell innovation to reviewers; instead, they should emphasize innovation where it exists and indicate how it complements the proposal. Traditional methods can be used together in such a way to demonstrate an innovative perspective.

All ideas for research applications do not have to be high in novelty, but ideas should, when enacted through a research plan, have a significant impact on advancing science. Sometimes the steps forward are incremental in nature, but that is acceptable if it can be shown the early incremental steps will later result in larger steps, or a novel understanding.

On occasion, applicants/investigators want to explore an idea that is not novel to the literature but for which there is little literature support, or that is surrounded by controversial literature. This situation can be used by applicants/investigators to promote the need for replication and verification or to promote the need to resolve a controversy. This approach to innovation requires an extensive knowledge of the relevant literature.

WRITING THE APPROACH SECTION: GENERAL ISSUES

The Approach section is where the description and explanation of methods as they support the individual aims take place. The aims are the plans for research, and the approach gives life to the plans. In the Approach section applicants should present pilot data as well as the statistical design and proposed analyses. The key to this section is applicant organization, clarity of writing, and enhancement of the aims with data and methods that will assure the reviewers that the application will have an impact on the field of science.

Reviewing the Aims section before writing about the aims in the Methodology section can help applicants/investigators avoid aims drift. As discussed in Chapter 5, aims drift occurs when the description of the approach and methods seems to result in secondary aims that evolve out of the methods and design but are not part of the initial aims or supported by the premise.

Often there are additional analyses that can be performed with a data set. Sometimes these additional analyses can be discussed in the Analysis section, but if they do not fit with the initial aims then they can be mentioned in the Analysis section as areas for future research. Sometimes aims drift occurs because a review of the aims and the justification of those aims reveal another direction. If investigators find they have aims drift or additional analyses they could conduct, they should rethink what the aims should be. Reviewers will focus on the length of time they think is required to complete each aim. If investigator outcomes are to be efficient, analyses must be completed within the proposed number of years of funding. Researchers must convince reviewers in the Approach section that the proposed aims will be completed with the proposed scope of time and budget.

Chapter 5 addressed how researchers identifying aims drift differs from proposing an exploratory aim. The term *exploratory* can emerge from the development of the methods and the support studies mentioned in the Approach section after the applicant has reviewed the evidence to support an aim. An exploratory aim proposed by applicants that is not well supported by the literature or preliminary data may not be perceived as strong by Study Section reviewers or at the program funding level. "Exploratory" can mean (a) elevated risk, (b) an aim that is not well supported by literature or pilot data, (c) an aim of less importance but still worth examining, or (d) an aim that would gather data to support the next step in the research process. Applicants must show that an exploratory aim fits within a context that supports all the aims. Often investigators think that three aims as a package is a goal for 5 years of funding for an R01 grant. Four years of funding for two solid aims would be a better approach, and two aims are acceptable at the funding level. A weak aim could result in a poor score for the overall proposal.

Sometimes the aims lead to procedures that have not been validated. Applicants/investigators may need to validate procedures that they did not establish before developing the aims of the proposal. Applicants/investigators should not promote procedures as aims if failure to establish the procedure would make the remainder of the aims nonfunctional. Applicants/investigators can provide assurances that they will rigorously establish procedures. If a procedure is highly novel and useful to the field of science, then a proposal to show the rigor and value of the procedure might be considered before moving on to the larger application. Sometimes, System 2 can provide funds to help establish a procedure that will support a larger research proposal. Applicants/investigators should explore patenting a highly novel procedure before exposing it to the public or in a general review.

There can be a fine line between detail and complexity in the Approach section. As discussed in Chapter 4, sentences that are long, refer to previous statements for clarification, or have indefinite referents (e.g., "this procedure . . .," which could refer to more than one procedure), or words with multiple meanings, such as "significant," make the reviewers work hard and can overload working memory.

For reviewers, the time needed for applicants/investigators to complete the procedures and the analyses is a question that keeps building with the description of each experiment. This is particularly true if applicants/investigators present methodology in a complex or dense manner. A common statement from reviewers as applied to complex proposals is, "It appears that the proposed recruitment and assessment of subjects or material elements is ambitious for the proposed time frame and the proposed budget." Early and or new investigators are more likely to receive this comment than established investigators because judging the ability of a budget and time frame to support aims and methods requires experience.

If reviewers find the timeline for the completion of aims is addressed briefly as an opening paragraph in the Approach section, this provides them with a feeling of assurance that the timeline for initiation and completion is part of the planning process. If the timeline presentation by applicants/investigators is evident, the reviewers can make this clear when presenting a study to the review panel.

Experienced investigators understand that grant-funded research is time limited with respect to the budget, even though much of the work of publication, data sharing, and thinking about the outcomes is done after the budget has been expended. Experienced grantees often build in time for these activities in the last year of a budget period. If completion of data gathering is planned to take all the applicants'/investigators' budgeted time, then there will be a delay in them resubmitting the application for renewal or pursuing hypotheses supported by the newly collected data.

WRITING THE APPROACH SECTION: COMMON STATISTICAL ISSUES

Statistical analyses should be treated like aims; that is, the proposed analyses should be focused, integrated, and promoted for what they can accomplish with a specific data set. Sample size and data distributions are two common conditions that strengthen or weaken the value of analytic approaches. The more evidence applicants/investigators can provide to support the assumptions of the analytic procedures, the stronger the application is. If there are alternative analyses that can be considered, those should be discussed.

Issues of reliability and reproducibility have shaped much of the recent policy at NIH and other funding agencies. Sample size, and how to estimate it, has always been important and may require more than just a cursory discussion because sample size determines not only the power to detect specific outcomes but also the effort needed for subject recruitment. NIH now offers a tool to compute power to not only promote power estimates but also to unify the method for obtaining estimates and the expectation for reporting (see https://researchmethodsresources.nih.gov/grt-calculator).

Providing the effect sizes for cited research can be persuasive if those effect sizes are of reasonable size for the proposed methods and outcomes. Investigators reporting just the p value significance of outcomes do not provide staunch support of a premise if effect sizes are not reported as well. Reporting the actual value of a correlation in a study (i.e., $r[df] = .66$) is more convincing than reporting simply that correlations were significant (i.e., that an r was significant at $p < .05$) or that previous studies have reported a significant outcome but the specific statistical data from those studies are not provided.

In Chapter 5, I suggested that exploring extra aims might be interesting but that without a strong reason for exploring them the extra aims might be considered as weak. The same can be said of suggesting extra analyses because the analyses might be interesting. Once the main analytic plan has been discussed and promoted, do not suggest after-thought or clean-up analyses that have not been well thought out. "Well thought out" means that power estimates and distribution of data have been estimated.

THE ENVIRONMENT AND INVESTIGATOR SECTIONS ARE "SHOW-OFF" SECTIONS

Chapter 8 discusses how federal funding agencies are under pressure to somehow mask the information about investigators and the environment until the main body of the proposal has been reviewed and rated. This is an attempt to reduce inequities in research funding due to the reputation of the System 1 investigator or the System 2 supporting institution. These issues are further addressed in Chapter 10.

The Environment section is where the quality of the System 2 support becomes especially important for System 1 applicants/investigators to show that they are working within a System 2 that can support the approach to include professional colleagues, statistical support, and technical support/ skill and recruitment of subjects. Administrative letters of support are critical to show access to space and time as well as any required methodology or technology. The presence of centers, special programs, training resources, and special collaborations can all be promoted by applicants/investigators as potential collateral support for a research project and the investigators supported by that project. The supporting environment description is critical to training grants (see Chapter 9). The capability of the environment to support animal or other nonhuman studies is important to those studies (letters showing access to support are especially important). Environments that are proposed by applicants/investigators to provide human subjects recruitment opportunities require careful description and letters of support showing access to those environments and their resources for recruiting subjects. The ability to recruit human subjects within a specified time period is always a critical component of clinical trials.

Reviewer evaluation of the Investigator section relies not only on the applicant's/investigator's description of activities relative to the aims, methods, and goals of the study but also, very heavily, on the match of the biographical sketches to the aims and methods of the proposed research. It is important that the biographical sketches be tailored as much as possible to the specific Aims and Approach sections of the submitted application. Often, Co-Principal Investigators, or important contributing investigators, will provide a biographical sketch that is not tailored to the proposal or updated, and that might lower the enthusiasm for reviewers. It is advisable to have someone from the research team or supporting System 2 (e.g., the grant support workforce [GSWF]) collect the biological sketches and examine them for pertinence to the proposed research. These forms are subject to frequent change as well (see National Institute of Allergy and Infectious Diseases, 2021), and the biographical sketch format used the last submission may not be correct if the rules for writing it have changed.

Good teams have to work together to be competitive, and the organizational components of the application should be clarified. It is not enough to see the biographical sketches of the team members; applicants/investigators should provide some discussion of how they have worked together in the past, or will work together in the future, to achieve the goals of the aims. The reviewers should be able to comment on the functional organization of the research.

Reviewers carefully examine the time an applicant/investigator proposes to devote to a funded application. The time they allot for a project must be credible for the proposed project and plausible in light of their ongoing activities and responsibilities. Members of the research team who are contracting to provide specific services for a limited period of time (e.g., biological assay, specified subject population recruitment) can be considered consultants. Consultants should always provide a letter of support for a specific role in the proposed research and indicate how much time is expected to be expended within that role. Budgets must show support for the presence of investigators and consultants. Unpaid consultants and investigators make reviewers and Program Officers (POs) skeptical.

KEY TAKEAWAYS FROM CHAPTER 6

- The Significance section promotes the overall research goals provided in the Aims section. This section is where an applicant/investigator states the evidence they have that the aims and hypotheses are worth pursuing. The evidence for one's case becomes stronger with support from the literature, proof-of-concept publications, and pilot data.

- Biological variables, such as sex, age, and any other biological factor significant to advancing the aims should be considered in the Significance section. The subject inclusion/exclusion and recruitment plans are part of the scientific potential and replicability of the aims outcomes.

- The Innovation section can be considered an extension of the Significance section but dwells on what is new in the Aims section or in the methods to support the aims. This section is where applicants/investigators need to consider what new information will be produced and what has to change in their perspective or approach to produce that new information. Less innovative studies can still move the field forward or provide a better foundation for moving forward.

- The Approach section is where all the power that has been packed into the Aims, Significance, and Innovation sections enters the freeway and moves the reviewers through the methods for each aim with power and skill. If the writing is too dense here, then one can get stuck, and all the enthusiasm generated for the reviewers in the previous sections stalls in a methods traffic jam. Keeping on the track outlined by the original aims is very important, and the potential for finishing the projects should be emphasized.

- The keys to competing well within the Approach section of a grant application are applicant organization, clarity of writing, and enhancement of the aims with data and methods that will assure the reviewers that the application will have an impact on the field of science.

- The analytic approach should be promoted, reasoned, on target for each aim, and well thought out with respect to the outcomes of every analytic procedure.

- Support from System 2 institutions should be made clear with respect to the resources needed to start and complete the investigation of the aims.

- The biographical sketch supports the basis for the proposed application of skill outlined in the methods and should be tailored to the aims and methods. The applicant's/investigator's skills promoted in the biographical sketches should surface in the Approach section in a discussion of an organized team approach to carrying out the aims.

- The aims, as supported by the approach, should produce data and outcomes that will need to be managed and potentially curated and provided for others. The plan for this should follow the funding agency's policies.

NOTICES

- **NOT-OD-18-109 Revision:** The NIH Announces Additional Review Criteria for Career Development Award Applications Involving Clinical Trials (https://grants.nih.gov/grants/guide/notice-files/NOT-OD-18-109.html)

- **NOT-OD-17-118:** The NIH Announces New Review Criteria for Research Project Applications Involving Clinical Trials (https://grants.nih.gov/grants/guide/notice-files/NOT-OD-17-118.html)

- **NOT-OD-17-121:** RESCINDED: The NIH Announces New Review Criteria for Career Development Award Applications Involving Clinical Trials (https://grants.nih.gov/grants/guide/notice-files/NOT-OD-17-121.html) This was replaced by NOT-OD-18-109

- **NOT-OD-17-122:** The NIH Announces New Review Criteria for Ruth L. Kirschstein National Research Service Award (NRSA) Individual Fellowship Applications Involving Research Experiences in Clinical Trials (https://grants.nih.gov/grants/guide/notice-files/NOT-OD-17-122.html)

- **NOT-OD-17-123:** The NIH Announces New Review Criteria for Ruth L. Kirschstein National Research Service Award (NRSA) Institutional Research Training Grants Involving Research Experiences in Clinical Trials (https://grants.nih.gov/grants/guide/notice-files/NOT-OD-17-123.html)

PART **III** INFLUENCING AND RESPONDING TO REVIEWER FEEDBACK

7 UNDERSTANDING AND WRITING FOR THE REVIEW PROCESS

Making the review process more understandable helps potential System 1/ System 2 grant applicants prepare better applications and makes the review process less mysterious and perhaps less anxiety provoking. The goal of this chapter is to demystify the review process by

- discussing the similarities and differences in grant review for the National Institutes of Health (NIH), National Science Foundation (NSF), and Department of Defense (DOD/DoD); most funding sources will follow a version of the NIH, NSF, and DoD review process;

- discussing factors in a review that are under the applicants'/investigators' control;

- examining factors that add variance to the review process and the outcomes of a review; and

- presenting the less well-known review processes.

https://doi.org/10.1037/0000390-008
Get Funded: A Practical Guide to Understanding the Grant Application Process and Writing Winning Proposals in the Behavioral and Biomedical Fields, by J. W. Elias

SIMILARITIES IN REVIEW PROCESSES

Most funding sources follow a similar process for review. The generic process is that a funding source (System 3) publishes a funding announcement that contains

- the goals of the announcement and the budget allotted,
- System 1 and System 2 applicant/investigator eligibility,
- the submission process,
- a timeline for submission,
- a timeline for review and funding, and
- a description of the review process.

A preliminary level of review is a referral process within the funding source that guides the application to a review group or returns the proposal to the applicants/investigators if the proposal does not adhere to submission guidelines and the goals of a funding announcement. If the funding source accepts the application for review, a review process takes place that most often involves peer review and feedback from that process.

Many funding sources have at least a two-tier review component. The first tier involves peer review by scientists or knowledgeable stakeholders who discuss the science and/or potential success of a proposal and provide written and scored feedback. The second tier consists of a funding stage where the funding source administration considers the value of the proposed research based on peer review and the goals and aims of the funding source. For some funding sources, there is a third tier to make certain the funding processes are guided by the review processes.

System 3 (the funding source) determines how peer reviewers will evaluate, score, and provide feedback to the funding source. Before sending peer reviews to the applicant/investigators, System 3 administrators will check the reviews for coherence with policy. Applicants typically receive written feedback and numerical scores as part of the peer review process. The scale of the feedback can vary and often depends on the financial status of the funding source. Feedback requires human capital and invites further interaction with applicants/investigators. Professional societies and research foundations that provide grant funding may not have sufficient sources to provide extensive feedback. Written feedback from the initial peer review process does not typically include the discussions held at higher levels of review. Applicants can request the perspective of Program Officers (POs).

Some funding sources have a multiple-step review process in which separate individuals serve in a chain of processing steps. The initial step in processing for federal resources is registering the System 2 supporting institution that will receive and oversee the on-site budget for the System 1 applicants/ investigators. System 1 applicants/investigators must register with System 3 to submit applications electronically through a System 3 portal. Federal funding agencies use a referral process that assigns applications to the correct funding component.

A BRIEF INTRODUCTION TO THE REVIEW PROCESSES OF THREE WELL-KNOWN FUNDING SOURCES: NIH, NSF, AND DoD/DOD

A brief discussion of how the federal agencies NIH, NSF, and DoD/DOD review grant applications will provide an example of how the generic review process blends with review processes specific to a funding agency.

NIH conducts research funding within its institutes and centers. NSF conducts research funding within directorates, and DoD/DOD conducts funding within programs. At each funding source, there are processes to make certain the applications are appropriate for an agency Funding Opportunity Announcement/ Notice of Funding Opportunity (FOA/NOFO) and are submitted according to the FOA/NOFO policy.

NIH Review Process

NIH provides extensive discussions of the grant application submission process on their website (see Chapter 2). At NIH, Scientific Review Officers (SROs) are charged with managing the review process. The SROs recruit reviewers, make review assignments, conduct a pre-meeting information session, and possibly a reviewer training session, conduct the review meeting discussion, oversee the scoring process, and ask for discussion of human subjects and budget concerns. Peer reviewers for NIH are to avoid discussions of funding per se but can comment on the adequacy of a requested budget, including the time requested to complete the proposed research. After the peer review meeting, the SRO provides a Summary Statement (see Chapter 8).

NIH discourages discussion of funding in peer review and by peer reviewers. NIH SROs, other than by writing a Summary Statement of the peer discussion, are not involved in the funding or recommendation process. The roster of NIH standing Study Sections can be viewed online at https://public.csr.nih.gov/ StudySections. Some of the reviewers will be permanent members who are appointed for specific terms of service. NIH also has ad hoc Study Sections that are recruited for special reviews, such as Study Section member conflicts or re-reviews of applications outside of the regular Study Section. Study Section members cannot have their own applications reviewed in the section in which they are permanent members. The names of the reviewers for these ad hoc sections are public knowledge, and their identity is published as part of Study Section rosters, but they may not be shown as assigned to a particular ad hoc panel. The Center for Scientific Review (CSR) reviews approximately 75% of the submitted applications. Most NIH institutes have a small number of SROs who put together review panels for special projects, large-budget applications, and training applications. The role of the NIH institute SRO is the same as it is in the CSR, and the review process is the same as in the CSR, but the review panels might be smaller.

Peer reviewers are SRO vetted and self-vetted for conflicts of interest with System 1 and System 2 submissions. Groups of reviewers assigned to a Study Section are asked to participate in a virtual meeting before the review process to discuss new policies and how the applications should be reviewed and scored. Final scoring (score range = 1–9; 1 is the best score) and final discussions are completed in a virtual or in-person meeting that may last 1-1/2 to 2 days. The peer review process allows the earliest time from submission to funding to be 10 months.

Funding is discussed in meetings at a second level of review by POs and institute directors. Funding is based on peer review scores, the Summary Statement, portfolio needs, and goals of the NIH institute or center. The initial program funding meetings produce a slate of potential candidates for funding. This slate is presented to NIH institute council members for comment on quality and procedures.

The slate of program-approved applications is reviewed at NIH council meetings. Council members consist of successful scientists and science stakeholders who represent the general and science community. Each institute has its own council meetings that follow the review cycle. The NIH has an NIH council as well. Some NIH institutes publish a funding line that is based on scores when the budget has been determined for the year.

NSF Review Process

NSF accepts targeted program solicitations and some unsolicited investigator-driven submissions. The NSF Proposal and Award Policies and Procedures Guide describes the submission and review process (https://www.nsf.gov/pubs/policydocs/pappg22_1/pappg_6.jsp). Within this resource, NSF provides a succinct illustration of its grant review and management processes (https://www.nsf.gov/pubs/policydocs/pappg18_1/pappg_3ex1.pdf). NSF also evaluates their grant applications for their contributions to the broader aspects of society and communities. The broader aspect of the society and community components distinguishes NSF review criteria from other funding agency review criteria and is a part of applications that should be well developed. The NSF SRO/PO can give insight into this component of NSF grant proposals.

NSF proposals are reviewed in study sections by at least three reviewers who are assigned to an application by a PO. Reviews can be submitted by email (ad hoc review) or discussed in person or at virtual meetings. The reviewers are independent of NSF and are vetted for conflicts of interest with the submitting System 1 or System 2. The reviewers provide written feedback and score evaluation. The scale for NSF scoring is 1 to 5, with 5 being the best score. NSF POs, when compared with NIH POs, have a combined role as SRO and PO. NSF SRO/POs choose the reviewers; manage the review process; write feedback commentary; and make a recommendation for funding to an NSF division director, who makes a funding decision and passes that decision on to the Division of Grants and Agreements. There are no standing Study Sections, but reviewers can evaluate a proposal multiple times. Many SROs/POs at NSF are recruited from the active scientist community and serve for up to 3 years in the SRO/PO position. Training is required upon entry into the NSF SRO/PO role. There are no funding lines because the scores are part of a weighted process that includes the judgment of the PO and the NSF directors. NSF states it would like to have the time from submission to funding be 6 to 7 months, but the funding process itself adds to this timeframe expectation for review and funding.

DoD/DOD Review Process

DoD/DOD offers a Congressionally Directed Medical Research Program (CDMRP) whereby biomedical applications are carefully reviewed for compliance with an FOA/NOFO and may require an initial submission of a white paper describing the basic proposal. The white paper is reviewed by DoD/DOD POs for its match to a Program Announcement (PA) and DoD/DOD goals. DoD/DOD describes the review process as "two tiered." In the first tier, there are review panels that are tailored to a specific PA, but there are no standing panels, and there is no public list of reviewers tied to a specific review panel.

DoD/DOD publishes a list of review participants once a year. Peer review for an application is described as consisting of two or more scientists and a consumer–stakeholder. A written Summary Statement of the evaluation of strengths and weaknesses is provided by DoD/DOD contractors.

DoD/DOD describes the second tier of evaluation in this way: "Programmatic review, the second tier of review, is a comparison-based process in which applications of high scientific and technical merit from the entire array of disciplines and specialty areas compete in a common pool" (https://cdmrp.health.mil/about/2tierRevProcess).

The Programmatic Panel tier of review relies on members who are chosen for their expertise in science areas, policy, the military, and specialty areas consistent with review needs. A key part of the description of the second tier of review is that the panel is directed to make comparisons of applications to see which will become the higher ranked applications in the pool. A list of Programmatic Panel members is published every year. Applications recommended for funding at the second tier of review are approved by the Commanding General, U.S. Army Medical Research and Development Command, and the DoD/DOD health agency. There is an Inquiry Review Panel (IRP) process for applicants who believe that reviewers made factual or procedural errors. The CDMRP IRP can address concerns about peer review or program review.

The DoD/DOD funding cycle can take up to 2 years, which is longer than the time required from submission to funding for many funding sources (see https://cdmrp.health.mil/about/2tierRevProcess). There are no standing review sections and no publicly published funding lines.

The Review Process for Other Federal Funding Agencies

At the Grants.Gov (https://www.grants.gov) site, potential applicants/investigators can explore federal grant funding sources and their review process. Applicants/investigators should carefully check the dates of announcements and the source of the information. Websites are frequently updated, and links may be broken. Individual users' computer security systems will alert them to some suspect websites.

REVIEW COMPONENTS THAT ADD PREDICTABILITY TO THE REVIEW PROCESS

The policies and procedures that guide the review process are predictable and dependable. When there are changes in policies and procedures, these are announced and, in terms of submission processes, tend to be incremental changes. The predictable components of the review process are discussed in the following list:

1. System 3 funding agencies publish application due dates (standard dates and FOA/NOFO and Research Funding Announcement [RFA] specific dates) online to allow for clear submission planning deadlines. The online publication of standard due dates also provides (a) approximate Study Section

review dates, (b) approximate council review dates, and (c) approximate funding dates. If agency referral personnel clear an application for a funding agency review, the approximate timeline for the remaining review-related processes is easily calculated.

Applicants who are not prepared for an upcoming submission date can check the FOA/NOFO expiration dates, which indicate how long a particular FOA/NOFO has been accepting applications and how many more cycles or submission dates are available. Funding agencies may renew FOAs/NOFOs before or after they expire, but applicants cannot count on a renewal when planning to submit at the last due date. Sometimes POs can provide FOA/NOFO renewal information. More complicated grant applications, like center grants, will have fewer submission dates within a particular budget year.

2. Applicants are in control of submitting applications accurately by virtue of filling out the application forms appropriately and by following policy regarding how to complete the forms. The forms and preapplication forms can be complicated, and frustrating but accurate completion is the responsibility of the applicant. Often, applicants have concerns about whether the information requested—for example, letters of support—should be submitted if the policy guiding the submission is not clear or seems superfluous (e.g., training grants that require too many support letters). Applicants can request clarification of submission policies from the SRO or PO noted in the FOA/NOFO. Before submitting their proposals, applicants can address concerns (depending on the issue) to the funding institute contact listed in the announcement.

Applicants should be aware that the funding agency contacts listed in the FOA/NOFO may change after an application is assigned for review or funding responsibility. Applicants should check the funding agency submissions portal for the appropriate agency contacts. NIH, for example, uses the Electronic Research Administration (eRA) Commons (https://www.era.nih.gov/) for communications with applicants.

3. System 1 applicants can gain control over the System 2 presubmission processes if they contact the appropriate System 2 officials (sponsored programs, offices of research, offices of the dean) to alert System 2 that a grant application is coming through the system from System 1 researchers. Listed System 2 contacts may be available over a window of time, but not necessarily within a short window before submission.

Contacting System 2 officials or administration staff about submission issues only a few days or hours before the deadline may result in no response and no sign-off for a submission. System 2 administrators take vacations and holidays, travel, have family emergencies, get sick, and run out of time, just like applicants. Applicants should give the administration at least a week's notice that they will need a sign-off at a certain time. Many administrators, if provided with advance notice, can designate a signee. The budget of the grant, and its scope, are what administration staff like to examine, in particular, if there is any new laboratory space, construction, or installations required.

4. Applicants have variable control over the System 3 funding institute assignment by virtue of their topic and sometimes by request upon submission. Applicants can request a particular institute or Study Section upon submission. System 3 funding referral personnel route applications to the appropriate Study Section and must approve the request for an initial assignment or reassignment. NIH applicants may be surprised to see that the referral process results in a primary institute as well as secondary institutes for funding. The institute-claiming process is described in the next list item.

    When notice of assignment to peer review has been received, applicants should check to see whether their expectations have been met. Sometimes applicants are surprised to see that their application is being reviewed by a within-NIH institute Study Section and not a CSR study section.

5. Some applications on topics encouraged by funding sources can get variable reviews in Study Sections not specifically focused on those topics. To concentrate expertise on a specific topic area (e.g., health literacy), NIH develops PAs that have specified criteria (i.e., PAs with Special Review Criteria and/or Specific Receipt Dates [PARs]). PARS, as indicated by their title, have a special receipt date and review considerations, such as focused expertise on the part of reviewers who value that area of research. Applicants should look for these opportunities when reviewers from standing Study Sections who are focused on more common issues for that Study Section indicate that an applicant/investigator's topic is of little significance. When NSF offers (PAs) that are highly focused, this reduces the potential for reviewers to devalue the topic. The same is true of other funding agencies that offer programs that are highly mission driven.

6. Although not thought of as reducing variability in scoring, the point of discussing applications in a peer review panel is to develop a reasoned consensus on the reviewers' perspective. Seeking a reasonable consensus among reviewers ideally has the effect of reducing variability in individual scoring. Most panel members will vote within the scoring range of the assigned reviewers, or very close to it.

7. To stabilize scoring, for some applications (e.g., R01 applications), the NIH review groups, including grants reviewed within the NIH institutes and the CSR, average the scores over the number of applications discussed within a Study Section for the past three meetings. The scores are multiplied by 10 to establish a mean and standard deviation for the Study Section meeting. The CSR administration converts the scores to percentiles for presentation. The point of converting scores to percentiles is to account for Study Sections that, on average, provide larger or smaller spreads of scores or when several strong applications within a peer review round could skew scores toward the lower end (better scores; see https://www.niaid.nih.gov/grants-contracts/understand-paylines-percentiles).

8. The primary source of control of the review process by applicants is the quality and presentation of the science in an application. Does the application meet the goals of the FOA/NOFO and the mission of the funding source?

VARIABLES THAT ADD TO THE COMPLEXITY AND MYSTIQUE OF SYSTEM 3 REVIEW

Peer review is the backbone of science with respect to criteria for publication, career advancement, awards, and certainly grant funding. The general concept of peer reviewing is rather well known and relatively simple. Peer review as a functioning process is complex, and some processes add to the variability of outcomes.

To gain perspective on these sources of variability in review outcomes, one should consider that not all the sources of variability discussed in this chapter are operating for each submission, and they do not make peer review invalid. Review policies are designed to provide a similar process for each review panel, but the conditions for each Study Section will change, and there are various conditions that are inherent to peer review. When humans and technology interact to produce a product within a brief amount of time, there will be noise in the system.

Major changes to the review process are rare, but NIH recently announced potential changes that will place less emphasis on the reputation of System 3 institutions. The proposed change in the focus of the review process and the reviewers' directions for evaluating applications should affect almost all of the NIH activity codes. A special emphasis is placed on changing the review focus for fellowship applications (see Lauer & Byrnes, 2023).

The proposed changes were made in response to criticism that the top-funded institutions continue to receive the majority of funding and that this trend might represent bias toward reputation and not the quality of an application with respect to significance, feasibility of method, and potential for completion of a project as designed. As further discussed in Chapter 8, in the NIH Summary Statement some of the proposed changes in review focus will relieve reviewers of the need to evaluate how applicants are responding to the administrative requirements in a proposal, such as evaluating data management plans (see Byrnes, 2022).

Processes and issues that add to the variability of outcomes include the following:

- The NIH review referral processes assign applications on the basis of Study Section expertise. The goal of all funding agencies is to find the appropriate expertise for the science represented in an FOA/NOFO. Funding agencies cannot predict how many applications of a particular kind will be submitted and assigned to a specific Study Section or SRO. For example, a Study Section that would normally receive eight applications on the topic of depression might receive 15, resulting in the assignment of more applications to Study Section members with expertise in depression, or recruiting professionals with additional expertise on depression, or spreading expertise among the existing panel members.

- Finding reviewers who perfectly match the designated scientific coverage of a Study Section can be difficult. SROs do their best to recruit reviewers who match the science, but they must compete for reviewers from other Study Sections and other sources. As life for everyone becomes more complicated, it can be more difficult for reviewers to find the time to provide proper service to a Study Section. RFAs require that SROs put together a review group just for that RFA, and the SRO often does not have a core of reviewers to whom they can assign applications while searching for additional reviewers. SROs for standing sections at NIH have a core of reviewers on whom they can usually count to attend a review meeting.

- A breakdown of the professional experience status of reviewers from 2015 to 2020 can be viewed at https://public.csr.nih.gov/AboutCSR/Evaluations# reviewer_demographics. NIH has made an effort to recruit applicants who are in the early stages of their careers. Early Career Reviewers (ECRs) must show evidence of active research in the form of publications and have applied to NIH and received a Summary Statement, but they need not have been funded for an R01. ECRs constitute only a small percentage of reviewers.

- It can be difficult to assign reviewers who have equal knowledge of the science and the methods in an application. There are typically three reviewers assigned to a submitted application. The primary (first) and secondary (second) reviewers usually have the scientific expertise required to evaluate an application. An assigned third reviewer may have expertise similar to the first and second reviewers or may have proficiency in design and analysis or of a narrower component of the science in an application.

- It can be difficult to find reviewers for new and emerging areas of science. There is always a concern that specialized reviewers serve as gatekeepers for new approaches in innovative applications; that is, other panel members in a Study Section may defer to the expertise of the specialized reviewer. Deferring to an expert is a reasonable decision, but it would be better if expertise in a scientific area were dispersed across the review panel.

- Sometimes applications cannot all be reviewed in a Study Section meeting because of restrictions on human capital and time. This results in a premeeting triage process. This process and its significant effect on the Summary Statement are discussed in Chapter 8.

- There is always a churn of reviewers at Study Section meetings because the same reviewers do not attend each meeting. About 40% to 30% of reviewers will not be permanent members, and the permanent members may not attend every meeting. This can add variability to the scoring processes as new members become adjusted to the review style of each Study Section and its scoring milieu.

- One cannot predict on what day, and at what time of day, an application will be reviewed; these may vary because of fatigue or a need to move more quickly through the review of applications to meet time constraints.

- Reviewers will devote varying amounts of time to evaluating applications, and well-written applications can be more efficiently reviewed and scored. The number of components and social issues reviewers must attend to in applications has increased over the years. The proposed changes in CSR review (see Chapter 8) will address reviewer load issues.

- Not all review processes require the reviewing and the funding process to be separate functions. NIH has distinct roles for the SRO and PO, and at NSF the SRO and the PO roles overlap. At DoD/DOD, experienced and trained contractors manage much of the review process.

- Applicants may be aware of the roster of reviewers before a Study Section meeting and may be able to point out potential conflicts with reviewers, but applicants cannot select their reviewers and cannot attend the meeting in which their application is peer reviewed. This is necessary to protect the confidentiality of the reviewers. Applicants must wait for a Summary Statement to provide more information so they can, if they wish, rebut reviewers' comments.

- Applicants can contribute information to POs for funding meetings but cannot participate in program funding decisions or the supervising council discussions of proposals.

- System 3 funding sources with smaller budgets for research review, funding, and funded-grant maintenance may provide less elaborate review processes and feedback.

- System 3 sources of funding cannot fund all the applications they would like to because of budget constraints. Budgets are not predictable from year to year because the cost of research is always increasing, but there are special concerns (e.g., disasters, the COVID-19 pandemic) that System 3 sources set money aside for, and an increase in a System 3 budget can be offset by the number of applications received.

- As noted, because of restrictions in the resources of System 3 funding applicants do not have complete control of assignments to Study Sections, although they can request a specific assignment.

- As science changes, the focus of review sections can change. For example, the financial growth of the National Institute on Aging (NIA) and the overall NIH focus on biological variables, such as age, has resulted in more Study Sections receiving applications that required review proficiency in age-related issues.

- As policy issues work their way into the review process SROs, POs, and reviewers must educate themselves and learn the nuances of the policies and how those policies influence the review process.

- It is hard to get reviewers to spread their scores in review when they know funding lines are low or funding is tight; consequently, small differences in scores can make a big difference in funding. Peer review is not always a concise match between scoring and application quality.

- Special reviews for RFAs will have to find ad hoc members for the specific RFA review. The RFA reviews will likely be less cohesive and produce more varied scores. There likely is no opportunity to resubmit an application to an RFA. Applicants can, upon advice from the funding source, submit the idea for the RFA as a new application in the form of a PA.

- Despite all the reliable technology involved in uploading and downloading applications, and review of applications, there is the potential for error. SROs and reviewers should be certain that reviewers have received the correct application to assess and that they have uploaded reviews correctly for each separate application and have not duplicated reviews before or after a meeting.

THE EXPANDED ROLES OF THE SRO AND THE PO IN PEER REVIEW

One of the goals of becoming more grant literate and experienced as an applicant/investigator is to gain a better understanding of the peer review process and a more thorough knowledge of how to respond to that process. Experience with application submission and review is a good teacher, and yet, even if one has participated as a reviewer for a funding source it would still be difficult to have a complete understanding of a particular funding source's review process without having worked as an SRO in that system. The same can be said of how well applicants/investigators can understand the funding process without having worked at the PO level. The NSF SRO/PO combined position and the recruitment of researchers from the community for short-term positions as SROs/POs do allow that kind of working experience for potential applicants/investigators.

With some exceptions, the scores and reviews of a grant application determine the application's potential to receive funding. Grant funding is not capricious in nature, but it can at times appear mercurial if a program administration funds applications that are outside a stated funding line. Funding outside of a particular line gives the fortunate applicants/investigators faith in the review process and provides those who hear about such funding hope that it is guided by, but not locked into, the scores from peer review.

Although the term *SRO* is well known, the functioning and training of SROs are not. SROs are the official government representative for a meeting and the processes that take place within that meeting. This is a role that carries both authority over and responsibility to manage the meeting according to best practices and policy. The SRO has the authority to remove individuals from the official government meeting if they are disrupting it. The SROs are responsible for reviews of millions of dollars of proposed research, a responsibility they take very seriously.

SROs typically have a doctorate-level education or a medical doctor's degree. They have scientific experience and training in the areas of research for which they are administrators. Some have their own NIH laboratories that function within the NIH intramural process, but they do not compete for funding via the extramural process that involves grant application review. SROs receive constant training in the review process, including any updates on NIH policy and process. The CSR provides a description of the SRO role at https://public. csr.nih.gov/ForReviewers/MeetingOverview/RoleofSRO.

The SRO role is a busy and intense one, driven by the multiple review cycles, and it involves important interactions with reviewers before, during, and after the review meeting. The SRO role does not involve significant interaction with specific applicants unless there is a question involving materials to be added to an application or a policy issue to be discussed relative to the review process. Ironically, during meetings it may appear that the SRO is doing little beyond announcing the next review or break time, but everything that happens in the meeting is the responsibility of the SRO, including framing the comments for Summary Statements issued after the peer review meeting.

NIH appoints peer review meeting chairs to keep the review process moving according to the instructions of the SRO (which may be subtle). Meeting chairs are chosen by SROs for their ability to recognize when more information is needed or if a discussion has started wandering.

For most funding agencies, the SRO or PO position is considered a full-time career position. At some agencies, such as NSF, the SRO and PO roles are blended. Some NSF SRO/PO positions are designated "short-term" positions.

At NIH, the SRO is not involved in peer review, although they may monitor active peer reviews in meetings. It is the role of the PO to read the Summary Statement provided by the SRO and, if asked, communicate to the primary applicants/investigators what the summed and averaged peer review committee scores may mean relative to funding. It is also the role of the PO to discuss the scientific direction of a resubmission of a previous application.

The role of the PO within an institute is broad, and the training components for the PO may be ad hoc and experience related (e.g., in the form of on-the-job training). When providing advice, POs must think ahead about budgets and funding cycles. The PO's responsibilities extend to portfolio management (i.e., they monitor the progress of funded grants relative to the aims and timeline). The PO monitors the progress of a funded grant and reports back on this progress to the funding institute. POs promote science within their portfolios in an institute and are often the spokesperson for a grant that has produced a significant finding. The PO often has the vital role of representing an application at a council meeting, if needed, and represents an application at institute funding meetings. POs are expected to follow new scientific developments and promote those developments within their program and the portfolio of grants they personally supervise.

When a funding institute announces or implements policy changes, all the SRO and PO personnel within an agency are required to understand the changes and translate them for applicants/investigators. Like applicants/investigators, both SROs and POs have a learning curve for the application of new policies and their impact.

The term *cooperative agreement* was introduced in Chapter 2 as both a grant mechanism and a grant funding code. Applications submitted with a cooperative agreement grant funding code (e.g., U01 grants; see https://www.niaid.nih.gov/grants-contracts/cooperative-agreements) require and permit one or more POs to work in close conjunction with the grantees after funding.

The PO for a U grant can follow the course of the funded project more closely and provide rapid NIH assistance if needed. This cooperation requires the PO to schedule several meetings with the applicants/investigators in between the usual annual project reports. To avoid conflict of interest there may be two POs assigned to a cooperative agreement, with the duties split between budget management and scientific involvement.

NIH HAS A CLEAR FUNDING OVERSIGHT PROCESS CONSISTING OF APPOINTED COUNCIL MEMBERS

Each NIH funding agency or office has a council (see https://ofacp.cit.nih.gov/) that is formed from stakeholders in the community or scientists funded by the NIH institute. Serving as a council member for NIH is a busy honor because council members are involved with reviewing and approving new research initiatives, evaluating policy issues, and assessing science and funding trends within the institutes. The council member role requires significant task involvement with NIH before and after the brief—usually 2-day—council meetings. The public may request access to the public components of council meetings when they are in session (if space is available) or by live streaming. NIH provides post-meeting online access to the public portions of the meetings.

Council members typically meet with NIH officials three times a year. Council meetings are scheduled within a few weeks of the PO's receipt of reviews from Study Sections and the development of a slate of candidates for funding. The council members perform the important task of reviewing and approving the slate of applications that the NIH institutes submit for funding. A council may ask for a discussion of the applications, but the council members' role is not to rereview applications. Council members provide oversight by ensuring that applications are funded by policy and within the expected score ranges. Applications cannot be moved to a "funding list" until the council has approved that list, although applications on the council-approved slate may not be funded until after further discussion with the PO in a post-council funding meeting (see https://www.nigms.nih.gov/about/council/Pages/councilmeetingsandfunctions.aspx).

In the post-council funding meeting grants are chosen by the program administration to receive a Notice of Awards (NOA—often referred to as a *Notice of Grant Award* [NOGA]). Awards for that funding cycle will be forwarded to the Grants Management component of an institute, and the financial components of the application will be examined for responsiveness to policy, appropriate categories of funding, and suggestions for funding made during the peer review and by the funding institution (see the next section for further discussion of the Grants Management Officer [GMO] and the Grants Management Specialist [GMS]). Budget constraints in a funding cycle can result in an application being considered for funding in a later cycle. POs can advise applicants/investigators on this process.

THE ROLE OF GMOs AND GMSs

It is uncommon for applicants/investigators to interact with GMOs and GMSs before receiving funding. On occasion, there are budget issues during the planning stage of an application or during the preaward and postaward stages of grant funding that require contact by applicants/investigators with GMOs and GMSs. The NIH context of GMOs' and GMSs' work is described at https://grants.nih.gov/grants/pre-award-process.htm.

Knowledge of fiscal policy is a critical component of grantsmanship, although many applicants/investigators cede that knowledge to System 2 institutional administration. Just as would be expected in the private sector, grant applications that seek funds for the construction or upgrade of equipment may require legal validation of ownership and a description of any contractual obligations for services or equipment funded. Attention to this kind of detail by System 1 and System 2 applicants is paramount.

A GMO's workload is high, and it may be doubled when federal budgets are delayed or revised and when budget agreements for pending applications or previously funded grants must be redone. When GMOs provide clear deadlines for response, applicants/investigators should respond quickly. GMOs working within funding agencies review thousands of budgets; grantees, on the other hand, see a few budgets and may overestimate their own creativity. GMOs and GMSs do not need grant funds to maintain their existence: Applicants/investigators need grant funds. Do not commit fraud; do not falsify documents.

REVIEWING IS A COMPETITIVE PROCESS

When applicants/investigators begin writing a grant application they follow a set of rules to communicate information that will be distributed across several technological platforms until the application reaches the review stage. Applicant/investigator resources have an impact on the quality of the application. The System 3 resources available to support the review process and the pressure of limited funding define, in part, how peer reviewers perceive their task and how they will discuss and score the science in the application. The funding process is where the products of review have their impact. When there are limited System 3 resources for funding, the review process has more of an impact because small deviations in scores can make an application more or less competitive.

KEY TAKEAWAYS FROM CHAPTER 7

- Funding announcements determine the general scientific focus of an application and the processes for submission and review. The review process varies from System 3 funding source to System 3 funding source, but there is a great similarity in the general review process across funding sources.

- System 3 reviews require human and financial capital, and the funding sources with the largest budgets tend to have a more elaborate review and feedback process.

- The SRO role is an important one that encompasses wide-ranging responsibilities, including the recruitment and assignment of reviewers, ensuring reviewer education, serving as the federal representative at the review meeting, managing a peer review Study Section meeting, and writing a Summary Statement. NSF combines the role of the SRO and the PO and recruits 40% of the SRO/PO positions from the scientific community to serve short terms.

- There are a number of stabilizing factors in the review process that provide a regular progression and reduce variability. The most effective way for applicants/investigators to reduce variability in their personal review process is to provide a clear presentation of science in the application and to meet the goal of the FOA/NOFO and/or the mission of the funding source.

- There is inherent variability in reviewing that affects scoring because of the nature of the processes and because SROs recruit reviewers for their differences in scientific opinion and their expertise. Although the process of referral to a Study Section is based on the goodness-of-fit between the application and the reviewers' scientific expertise, an exact fit may not be possible.

- Peer review is where the competition of grant writing becomes apparent. Any inequities in an applicant's/investigator's ability to prepare an application for review or inequities in the review process could affect peer review and potential funding.

- The scientific community often characterizes the peer review Study Section meeting as the arena where the competition of grant writing takes place. Because of restrictions on resources, some applications are triaged by SROs and are not placed into a meeting for discussion and further competition.

- The funding process is where review outcomes have their impact as POs and System 3 administrations consider the science reflected in the reviews as presented in the vehicle for funding discussion—the Summary Statement. System 3 funding decisions are weighted by System 3 resources and goals as well as the products of review.

- The proposed changes for the NIH review process will likely become review policy at the end of the 2023 review cycles or at the beginning of the 2024 cycles. It will take some time to evaluate the effects of these changes on the funding process.

8

RESPONDING TO REVIEW FEEDBACK AND PREPARING FOR POTENTIAL RESUBMISSION

The Summary Statement is an instrument of communication and reformulation. As the primary form of feedback to applicants/investigators, it fits within a distributed-cognition model as it applies to grant writing. A *distributed-cognition system* has been described as one that has a component for both internal and external representations (Rogers & Ellis, 1994; Zhang & Patel, 2006). *Internal representations* are knowledge and structure in a person's mind, and *external representations* are knowledge and structure in the external environment. In a distributed-cognition model there is a potential for the information in a person's mind to be changed and framed by the external representations of knowledge required for exchange of information via technological transfer. The constraints of technology for grant submissions, and the use of forms designed by policy, frame the way ideas from applicants/investigators are transmitted to reviewers and then back again to the applicants/investigators in the form of a Summary Statement.

Figure 8.1 depicts the flow of communication from the internal representations of the minds of applicants/investigators to the minds of peer reviewers. The Summary Statement is the vehicle for transporting the internal representations of the peer reviewers, including a translation by a Scientific Review

https://doi.org/10.1037/0000390-009
Get Funded: A Practical Guide to Understanding the Grant Application Process and Writing Winning Proposals in the Behavioral and Biomedical Fields, by J. W. Elias

FIGURE 8.1. The Grant Application and Review Process Represented as Distributed Communication

Knowledge from the literature and data are translated to Aims, Significance, Innovation, and Approach proposed in a submitted application.

Technology artifacts = grant format, electronic submission, & preliminary review format.

Policy determines review guidelines for reviewers.

Reviews are provided within social network of study section

Applicant ideas are recast via summary statement feedback.

Officer (SRO), back to the applicants/investigators. As will be discussed, this is not one-to-one communication, and so the technological artifacts that support the communication between applicant and reviewer play a role in the communication process. A Program Announcement (PA) or Funding Opportunity Announcement/Notice of Funding Opportunity (FOA/NOFO) posted on a website as an initial communication among Systems 1, 2, and 3 is a system of communication made possible by, and shaped by, the communication systems developed by humans through the application of technology. Websites, grant application transmissions, and Summary Statements are human technological artifacts.

Applications are not communicated in real time to peer review panel members, so the information in the applications is packaged so that it conforms to the limitations of the distribution systems. Feedback to applicants/investigators from panel members is also formatted in a stylistic way. Applicants can receive feedback from reviewers in one of two forms:

- They may receive unfiltered direct feedback in the form of a review and accompanying scores to show evaluation relative to a scale and, potentially, other applications. This is the process for Pathway 1 triaged applications, discussed later in this chapter.

- Feedback can also come in the form of comments that are filtered through the comments of other primary reviewers, panel commentary, and an SRO's interpretation in the form of a Summary Statement. The panel deliberates to provide the scale evaluation, which can be further presented after the meeting in the form of a percentile.

As part of the feedback distribution system, communications allow for multiple readings of a research application. Electronic communication can permit more time for a reviewer to reflect on what a review should convey to the investigators. Discussions with Program Officers (POs) can occur in real time, although such communications are usually limited. The communication and feedback process in real time is different from a feedback process that occurs in the form of a formal Summary Statement with a set of scores. The higher the number of individuals who communicate in a chain of information distribution, the more opportunity there is for the original message to change.

As stated in Chapter 3, System 3 funding agencies, especially the National Institutes of Health (NIH), place a premium on seeing policy play out through the announcements that direct the writing process. The forms are designed to guide applicants'/investigators' writing toward the appropriate attention to the policy issues. When applicants see in an application a checklist, a specific form, or specific directions, they know, or should know, that these components represent policy. Peer reviewers are guided by their science expertise and experience, but these must mesh with the policy developed and expressed as required review components. Reviewers do not write reviews in free form but instead conform to a grant agency's policy, which means that information and ideas are being distributed between applicants and reviewers via specific formats that are dictated by that policy.

A single peer reviewer evaluates an electronic application according to a set process of some level of formality and conveys the rated basic values of the science application back through the information distribution system. Evaluation scores, and perhaps a percentile score, convey how the feedback compares with other reviews. This peer review cognitive-distribution process completes the transfer of information that began with the discussion of the application's original aims and hypotheses.

In the typical review process there is no live audition of an application with immediate feedback and discussion. It is likely that some individuals, including applicants/investigators, SROs, POs, and peer reviewers, communicate better across a distributed communication system than other means. Although what the impact is when Systems 1, 2, and 3 have different communication styles is not immediately clear, there likely are effects on the inclusion and interpretation of an application's content.

In Chapters 2 and 3, the grant application was presented as a product of policy that specifies, among other things, individual grant mechanisms and announcements that relate to grant-renewable status, applicant/investigator eligibility, FOA/NOFO–permissible goals, duration and levels of funding, and salary limitations. Policy-derived communication formats determine the specified times for submission and review as well as the forms and formats for submission. These determine how much time can be given to consideration of a written application, what can be written, what must be written, page limitations for what is written, and the rules for communicating ideas. Very few studies have examined how changes in communication policy affect the outcome of review.

Reviewers are instructed to form evaluative opinions about set issues and to report these opinions in specific ways (written comments, scores); on specific forms; under specific time constraints; and in different forums, including private reporting and discussion and group reporting and discussion. In line with the theory of distributed cognition, applicants'/investigators' initial ideas are not only relayed but also distributed across a complex network of policy-guided communication formats that transfer the original ideas across time and shape the reading context. Feedback from reviewers, review panel members, and the panel SRO is shaped by the same network of policy-guided communication formats. The format for grant application submissions likely has a trimming effect on original ideas, and the feedback from a review may suggest further trimming or alteration.

SUMMARY STATEMENTS REVEAL A CONGRUENCE OF THOUGHT BETWEEN REVIEWERS AND APPLICANTS

When applicants/investigators read the Summary Statement, they look for feedback on the ideas conveyed in the application. The reviewers have provided commentary on what they think the applicants/investigators are trying to convey.

The Summary Statement is not only an evaluation of an application, it also provides a degree of confirmation of the congruence between what was meant to be conveyed and what was conveyed. The technological artifacts of the review process do not allow misinterpretations to be resolved in the form of a Summary Statement. Response to a Summary Statement requires resubmission and further review using the same information-distribution tools and pathways.

Summary Statements that provide clear feedback on changes that could be made to improve the science in an application, or that clearly convey what the reviewer thought was lacking, are very helpful. When reviewers appear to have misperceived information in an application and applicants/investigators believe there is a lack of congruence between the meaning intended and the reviewer's understanding, this is very distressing to the applicants/investigators. If reviewers indicate that information is missing in an application and the information is in fact in the application, this is particularly disturbing to applicants/investigators. The Summary Statement, however, is not just a statement of understanding by reviewers, it is also an attempt to transform the science of the applicants.

TWO PATHWAYS TO A SUMMARY STATEMENT FROM NIH

The Summary Statement is one of the most important products of NIH and other funding agencies. System 3 resources result in a Summary Statement that is generated by means of two distributed pathways based on prereview meeting reviewer scores. Before the Study Section meeting, one distributed process is used to assign the applications and designate when reviews are needed. Upon receipt of the reviewers' scores, and a few days before the peer

FIGURE 8.2. Two Pathways for a Summary Statement

Pathway 1
Panel Peer Review

Assigned reviewer scores are in lower half of the pre-meeting score distribution. Application is discussed by panel in a study section meeting.

SRO summary of panel discussion included in a Summary Statement, plus panel Impact Scores, assigned reviewer comments and scores.

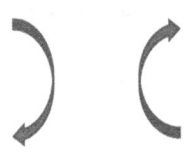

Pathway 2
Pre-meeting Review

Pre-meeting scores from assigned reviewers are in the upper half of the pre-meeting score distribution and the application is triaged and not discussed (ND).

No SRO summary and Summary Statement includes pre-meeting assigned reviewer comments and scores only.

Note. SRO = Scientific Review Officer.

review meeting, the Scientific Review Officer (SRO) will calculate the scores to designate applicants/investigators who scored in the lower and upper halves of the applications for that meeting. As shown in Figures 8.2 and 8.3, this process will effectively create two distinct pathways to receiving feedback. This upper half–lower half distinction is developed separately for the Early Stage Investigators (ESIs) and experienced investigators to provide a more

FIGURE 8.3. Pathway 1: Development of an Impact Score and SRO Summary of Discussion

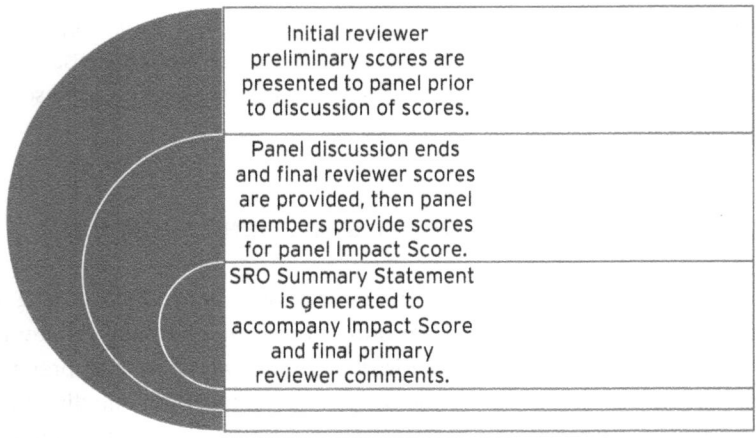

Initial reviewer preliminary scores are presented to panel prior to discussion of scores.

Panel discussion ends and final reviewer scores are provided, then panel members provide scores for panel Impact Score.

SRO Summary Statement is generated to accompany Impact Score and final primary reviewer comments.

Note. SRO = Scientific Review Officer.

equal review process for less experienced applicants, who are reviewed relative to and within their own group.

To establish the Summary Statement pathways, applications with the preliminary impact scores in the upper half of the distribution will be announced at the meeting, and if there are no requests by panelists or an original reviewer to bring an upper half application to review then the application follows Pathway 2.

Applicants who are designated to receive feedback via Pathway 1 receive (a) individual primary reviewer scores for each of the criteria, (b) written explanations of those criterion scores, (c) an overall Impact discussion from each primary reviewer, (d) an aggregate Impact Score generated from the scores of all the review panel members, and (e) an SRO summary of the discussion of the application. Applicants designated to receive feedback through Pathway 2 receive (a) premeeting primary reviewer scores for each of the criterion categories and (b) pre-meeting primary reviewer evaluations of each of the criteria. Pathway 2 applications will receive a "Not Discussed" (ND) notification on the Summary Statement followed by the above-noted ND reviews of the primary reviewers.

Assignment to Pathway 1 or Pathway 2 is a critical juncture in the review process because Pathway 1 applications are discussed in the review meeting and receive significantly more feedback in a Summary Statement that is based on a panel presentation and a panel review. Applicants who have scored in the upper half (ND) of the premeeting distribution of scores must respond to reviews that are not only more negative in tone but also likely more varied in perspective. Without a panel presentation, there is no opportunity for preliminary reviewers and a panel to reach a consensus on the accuracy and importance of the preliminary reviewers' concerns.

WHY IS THERE A PATHWAY 1 AND A PATHWAY 2?

The NIH Data Book and Center for Scientific Review (CSR) websites (https://report.nih.gov/nihdatabook/ and https://public.csr.nih.gov/, respectively) indicate that for 2020, there were an estimated 78,000 to 80,000 applications submitted as part of a slightly rising trend that began in 2015. There could be roughly 3 times that number of initial reviews provided for each application, resulting in roughly 240,000 individual reviews, approximately half of which are directed toward detailed Pathway 1 Summary Statements. Obviously, providing Summary Statements for that many applications in a timely fashion is a big operation that should be applauded for its efficiency.

Many funding agencies do not provide elaborate feedback like NIH does because they do not have the budget and the personnel.

Meeting time is precious, and reviewers find it difficult to maintain focus when meetings extend beyond 1-1/2 or 2 days. The cost to NIH balloons when there is an extra day of remuneration, including travel and hotel accommodations. If meetings averaged 2-1/2 days, the days on which meetings could start and end would be affected, and reviewers would be more likely to bargain to attend early or later days of the meetings. When panel members bargain to join or leave meetings early, the panels do not receive the full range of panel participation and scoring for each day. With each day of a meeting, the cost to reviewers and home institutions increases to accommodate for loss of instruction time and perhaps increased costs for child care and other life maintenance issues.

Applications with more concerns, such as those in Pathway 2, can take longer to review, and when it is obvious that an application has multiple concerns it is likely that, with the approval of the SRO, the chair of the meeting will curtail discussion and ask for final scores. Applications with better scores can then receive more attention. With an extra meeting day, the SRO and NIH support staff would be delayed one more day in providing scores and starting the feedback process. The load on the SRO and support staff to provide more complete Summary Statements would take a toll on human performance equity (bigger vs. smaller study sections) and efficiency. Given the constraints of finances, time versus performance, reviewer attention, and staff workload, the two-pathway review (although not officially acknowledged as such) is likely to continue as a necessary accommodation to an active research climate.

INTERPRETING THE IMPACT SCORE

The generation of the Impact Score or ND status is the focal point of the review process. The Impact Score signifies the potential for funding, for revision and resubmission, or for abandoning a set of aims and moving on to other research projects. The Impact Score for discussed applications is derived by averaging the final scores provided for an application from each of the panelists. Impact Scores and criterion scores range from 1 to 9 and are multiplied by 10. If an application is discussed in the meeting (Pathway 1), the in-person reported meeting scores from the assigned reviewers are used as guidelines for the panel members, who most likely do not read the application but listen to the discussion and, if necessary, ask for clarification of the points of discussion. Panelists can score outside the range of scores provided by the assigned reviewers, but they are typically asked to declare and defend that score in the meeting if the score is remarkably outside the initial reviewers' range. A panel member scoring outside the range may be asked to include a statement for the SRO that can be used in the Summary Statement. Because some mechanisms and Activity Codes (e.g., the R01) are typically reviewed by standing Study Sections, the scores are ranked to account for the variance in scoring among and within Study Sections and reported as percentiles derived by averaging across the last three meetings.

When applications are first received, panelists are instructed to search them for the inclusion of personnel with whom they may have a relationship that puts the panelist in conflict (e.g., a former student, a former mentor, someone with whom they are currently publishing together, the same or linked institutions). These panelists must temporarily leave the meeting when the conflicted application is discussed and do not score it. Impact Scores and ND status are released to the Electronic Research Administration (eRA) portal (https://grants. nih.gov/aboutoer/oer_offices/era.htm#:~:text=Electronic%20Research%20Administration%20(eRA)%20%7C,Health%20NIH%20Grants%20and%20Funding) typically 2 days after the meeting.

The meaning of the ND status is clear for that submission round. Applicants can resubmit an ND application as an A1 application or as a new application.

It is ironic that, for applicants who do receive a score, despite the elaborate review process, several months of work, and several months of waiting for a review, there is the likelihood that the meaning of the Impact Score relative to funding may not be clearly established until funding decisions are made. A decision on funding an application can be delayed over several review cycles until funding agencies make final decisions based on the available budget.

Scores at the 10th percentile or below should evoke a hopeful response from the applicant and the funding agency. Scores approaching the 20th percentile and slightly higher than the 20th percentile may still be in the fundable range depending on the primary institute's funding line and the funding institute's interest. (A funding line is sometimes established to indicate that an application with a certain score or percentile is likely to be funded.)

Many NIH institutes will publish funding lines that can be consulted, but some do not. It is always best to consult with the PO to see what a score might mean relative to the likelihood of receiving funding. For some mechanisms, such as fellowship and career awards, scores around 3 or slightly above can be competitive.

Pathway 2 applications do not receive an Impact Score and Summary Statement and are not eligible for funding. Some Pathway 1 applicants will receive Just-in-Time notifications to begin preparing the application for funding, but those notices are sent automatically to the top 30% of grants by score, and the funding line at each institute may not reach 30%.

Better scoring applications are often prepared by the SRO and released to the applicants and PO first until all the scored applications requiring a Summary Statement are completed. Summary Statements should start appearing within at least 1 week after the Study Section meeting.

THE WAIT AND THE INTERPRETIVE PROCESS

Applicants/investigators should look for a published funding line for the primary institute. To help the applicants understand scores relative to funding potential, many NIH institutes publish a funding line for specific Activity Codes (i.e., R01, R03, R21, K 24) that indicates what scores and/or percentiles are likely to be funded. These institutes and other federal funding agency funding lines are typically most reliable when they appear later in the funding year, when funding budgets by mechanism are most well established. A web search of published funding lines for any funding year will reveal that there is significant variance as to what scores are fundable among the 27 NIH institutes and centers that set funding lines. The funding lines vary by mechanism and, sometimes, by area of research (e.g., the National Institute on Aging [NIA] and funding lines for Alzheimer's disease research).

Applicants/investigators should also wait until they receive a Summary Statement and have reviewed it before contacting the primary institute's PO, who will not have a Summary Statement until it is posted. Once applicants/investigators have clear notes on issues raised in the review, they can contact the PO by email to set a time to discuss the application and its funding potential. The PO may already have had the opportunity to listen to the review; if not, they will have to wait to receive the Summary Statement and review the applicants'/investigators' notes before engaging in any discussion by phone, internet, or email. Note that the Summary Statements are constructed and edited by the SROs on the basis of the reviewers' in-meeting commentary and the written commentary.

The summary of the meeting discussion is provided to help applicants/investigators who are reviewed via Pathway 1 gain a better understanding of how the Impact Scores were influenced by meeting discussions.

- Some scores will suggest that the applicants/investigators wait for funding and prepare a rebuttal statement that the PO can use in a future meeting. The applicants/investigators can use the reviews to prepare this rebuttal. Depending on the interest of a secondary institute (often established before submission or review), a note to the secondary institute's PO might be useful. The secondary PO typically does not become involved in the funding process unless there is a reason for institutes to want to share funding. Nevertheless, by claiming the application in the initial referral process the secondary institute indicates interest.

- Some scores will be on the borderline between resubmission and waiting for funding without resubmission. In this case, applicants/investigators should use the reviews to consider what can be strengthened before resubmission. The required one-page rebuttal in an A1 resubmission is a useful tool to direct reviewers toward the issues of the application. Issues discussed by reviewers in peer review that applicants/investigators can clearly address will be to their advantage in a resubmission. If the initial submission is eventually deemed fundable, the later submission can be withdrawn. Once reviews have been provided and a Summary Statement issued, the previous reviews will be available to the next A1 (resubmission) review panel. For applications that are resubmitted as new ones, only the PO has access to the reviews from the initial application.

- Some scores will indicate that funding is not likely, and a resubmission should be prepared. Applicants/investigators will need to decide whether to (a) resubmit an A1 revised application that responds to the reviewers' comments with a direct rebuttal page and changes; or (b) resubmit a new application, where a rebuttal is not permitted but changes can be made, and the previous reviews will not be shown to the reviewers. The application likely will go back to the same panel unless a request to change panels is requested and permitted. New supportive data are of benefit, but the NIH page limitations will be in effect. When the concerns noted by reviewers require additional writing, some of the previous writing will need to be made more concise and precise.

- As noted, applications with an ND status will need to be resubmitted. These can be resubmitted as a new application. The benefit of submitting as a new application is that the review statements will not be provided to the next review panel (although the assigned reviewers may be the same). New pilot data may be of great benefit if there are questions regarding the benefit of the outcomes or the applicants'/investigators' ability to collect the data. Applicants/investigators who receive ND Pathway 2 reviews may want to consider changing or dropping aims with poor reviews or abandoning the project. They should refrain from rushing to make a resubmission until the reviews have had time to sink in and the emotional response has had some time to die down. It may be that the reviews, although unsettling, provide a clear direction for revision. This is more difficult with ND applications given that no consensus was reached via a panel review, but sometimes the reviews offer a direct pathway to correction that can be documented in an A1 resubmission.

SORTING POSITIVE AND NEGATIVE COMMENTARY FROM REVIEWERS

Once applicants/investigators have received the Summary Statement, it is useful to sort out the positive and negative comments into categories. One of the most important questions to ask about each comment is whether that comment reveals a science issue, a communication issue, or both.

The way the review criteria are structured focuses initially on the value of the aims (Aims, Significance, Innovation) and then on the potential for the aims to be adequately supported (Approach, Investigators, Institutional Support). Human subjects are not scored but are included in the potential for successful aims competition. Human subjects, and humane treatment of animals, is an area where a "bar-to-funding" code (an issue that needs to be resolved) can be attached to the application until the issues are addressed to the satisfaction of the funding agency grants management personnel, not the local institutional review board. The code is attached to the Summary Statement. Human subjects and inclusion categories (age, sex, minority status) should be considered as part of the potential for successful completion of the aims in the Aims section (see https://www.niaid.nih.gov/grants-contracts/human-subjects-inclusion-codes). The importance of designating a project as a clinical trial has augmented the importance of describing human subjects recruitment and protection as well as procedures for protecting and sharing the data collected.

This process of sorting helps focus the emotion that typically accompanies reviews of one's own work if the comments are mixed and/or suggest a resubmission. The suggested basic sorting categories include the following:

- Aims value—comments that question or approve of the development of the research aims

- Aims potential—comments related to successful completion of the aims, including the following:

  - comments that question or approve of the approach to producing/collecting and analyzing data that would support the aims; and

  - comments that question the researchers' capabilities or the institutional support or commitments; this category could also be perceived as addressing the applicant's/investigator's potential.

FIGURE 8.4. Organizational Scheme for Understanding Reviewer Comments

• Criterion score comments (+/–) related to:
   • Aims acceptance (structure, support, significance, innovation)
   • Aims potential for success (method recruitment, procedure, analysis, pilot data, investigator's potential, environmental support)

• Criterion score comments (+/–) related to:
   • Aims acceptance (structure, support, significance, innovation)
   • Aims potential for success (methods, recruitment, procedure, analysis, pilot data, investigator's potential, environmental support)

• Criterion score comments (+/–) related to:
   • Aims acceptance (structure, support, significance, innovation)
   • Aims potential for success (methods, recruitment, procedure, analysis, pilot data, investigator's potential, environmental support)

• Impact Score: (   ) SRO Summary +/– related to:
   • Aims acceptance (structure, support, significance, innovation)
   • Aims potential for success (methods, recruitment, procedure, analysis, pilot data, investigator's potential, environmental support, data management/sharing)

Note. SRO = Scientific Review Officer.

Figure 8.4 shows an organization scheme for sorting out scores and a review's primary emphasis.

COMMON QUESTIONS ABOUT THE VALUE OF RESEARCH AIMS

For applications submitted in response to a targeted FOA/NOFO, the aims may be provided in the FOA/NOFO, and there is limited choice regarding the general goal of the Aims section. In some cases, the general aims may be provided in the FOA/NOFO but there is some allowance for focusing on or trimming the aims. When the writing of the Aims section is directed by the FOA/NOFO, it is important to point out in that section, or in the Significance section, that the applicants are responding to a FOA/NOFO request that is shaping the choice of the aims.

Sometimes reviewers will overlook the direction of the FOA/NOFO even though it is available for reading. They are advised to read the FOA/NOFO to see whether it addresses particular aims. Reading and interpreting the FOA/NOFO is a significant amount of work for reviewers. Unless a Study Section is convened to focus on a specific FOA/NOFO, the reviewers may be reviewing applications with both targeted and investigator-initiated applications in the same Study Section. It is important to note, when reading and responding to the Summary Statement, whether the reviewers have ignored the directions of the FOA/NOFO. The reviewers, on the other hand, may point out that the applicants have ignored the directions of the FOA/NOFO, which is an unforced error on the part of the applicants.

The Aims section of an application is addressed first, so it is important that it receives thoughtful and helpful commentary. The aims per se do not receive a score. They are frequently commented on when reviewers discuss the Significance and Innovation sections. Comments in those sections could indicate that the applicant/investigator should provide better support for, or a better explanation of, the aims. A good starting point when reading the Summary Statement is to note the consensus of the panel with respect to the value and acceptance of the aims.

Disparities in scoring outcomes are common for applicants whose applications are not discussed in the Study Section meeting. Often, an ND status is due to large variance in reviewers' scoring. If there is a consensus among the reviewers across the review criteria in providing higher scores, then the application likely needs a complete reworking and rethinking.

When revising and resubmitting an application as an A1 application, it is important that the applicant/investigator point out to the next set of reviewers (some of whom may be the same as in the original panel) in the rebuttal page that there were disparities in scoring and that these issues have been specifically addressed. In many cases, disparate scores represent not only differences of opinion but also a lack of clear communication or persuasion on the part of the applicants/investigators. Disparities in scoring are annoying to applicants and tend to evoke anger, but a relevant question is, what can the applicants/investigators do to reduce the variance in scoring?

If a disparate score reflects inaccurate processing of information on the part of a reviewer, then this can be noted in a rebuttal letter requested by a PO for a funding meeting or in the rebuttal allowed for resubmitted applications. If the disparity is due to differences of opinion, then a more persuasive argument for that issue is needed in the resubmitted application. Often, the more negative reviewers are those who can provide the best insight into the perception of the aims and the approach. Tough but direct comments on what issues should be addressed, especially if they come with encouragement for revision, are useful.

It is not unusual for applicants/investigators to admit, after receiving reviews, that they "did not see that coming," meaning that there is another way to look at things they may have overlooked. This is not the same as being blindsided by a review that is indeed off the mark or clearly biased.

In areas of science that are newly developing, it is not unusual to submit multiple applications where a few "I did not anticipate that" comments occur in reaction to each Summary Statement. If applicants/investigators keep track of these statements and include the information in the next application, they may eventually refine the application to the point of receiving fundable scores.

WHAT TO DO WHEN THE AIMS, SIGNIFICANCE, AND INNOVATION RECEIVE NEGATIVE COMMENTS

A common cause of overall high scores is that the aims are not independent, to the point that a lack of support for an initial aim could compromise another aim or make it not worth completing. In Chapter 5, these were referred to as *cascading aims*. This makes the feasibility of the aims rather low for both the reviewers and the applicants/investigators.

Although it might seem reasonable to propose a series of experiments for which the success of the first experiment allows pursuit of the second, the funding for many Activity Codes is for the complete package of aims, and it is not easy to propose funding in stages unless the FOA/NOFO or grant Activity Code allows for it (e.g., an R34 Activity Code for planning grants). For some applications, the procedures required to complete an aim must be validated. These validation issues should be discussed in the Approach section, and it is up to the applicants/investigators to convince the review panel that the procedures will be validated. The need for such validation does not typically make the application more competitive.

In some cases, aims may be described as "dense," which is a signal that the writing in the short Aims section needs to be more on target and focused on primary issues, not the details of methodology. When aims are referred to as "ambitious," this suggests that the reviewers may think there are too many aims for a particular time period or the proposed budget. Such a comment indicates that the Summary Statement should be carefully searched for indications that the reviewers may be asking for more evidence of applicant/investigator expertise, an expansion of their expertise, or better pilot data.

When the scores for the Significance and Innovation components reach the 30 to 40 level, this indicates that the premise was not well supported or that the importance of the aims was considered moderately important, but not important enough to receive scores of 10 to 30. Scores for the Innovation section, in particular as they relate to the overall Impact Score, are hard to interpret. There are many interpretations of the Innovation section, and many peer reviewer comments on the NIH commentary sites reveal that Innovation is the least understood of the criterion scores. For example, the CSR recently suggested that the review of applications mix Innovation with Significance as a category for scoring. If Significance scores are excellent to very good and Innovation scores are good to moderate, then the issue might be that less innovative methods are acceptable. When the focus of a FOA/NOFO is on innovation, then moderate Innovation scores are not competitive.

If a review describes the aims as "underdeveloped," then there could be concerns that the brief Aims page has failed to support the importance of the aims; that there is a lack of coherence to the presentation of those aims; and that the reviewer was expecting more, or better, information. Such a comment could mean that the Significance section did not convincingly support the need for the aims or that the aims were not well supported by the methodology proposed in the Approach section. The reviewers ideally will fully describe what is behind their comments.

COMMON QUESTIONS ABOUT THE FOUNDATION OF THE METHODS PROPOSED TO SUPPORT THE AIMS

A review may indicate that the aims are found to be of quality but that the perceived potential for achieving the aims is moderate to weak. Feasibility is the issue when methods are directly challenged for the following reasons:

- appropriateness, or lack thereof;
- a poor defense or a lack of detail;
- weak support from pilot data;
- lack of statistical power or calculation;
- the potential for bias in subjects recruitment, or bias that is built into the procedures;
- questions about the applicant's/investigator's credentials for a proposed method or appropriate laboratory resources;
- lack of discussion of the rationale for selection of a method/analysis when there are multiple choices;
- lack of discussion of potential issues in methods and solutions for potential problems;
- an Approach section that reveals additional aims not identified in the initial Aims section, or the revelation of sub-aims; or

- statistical methods that are not specifically related to each aim and do not clearly reveal how they will support the aim.

The good news about most of these kinds of very specific concerns from reviewers is that they are mostly all fixable, and they can be directly addressed. When reviewers have concerns about the value of aims, then the arguments become more abstract, and the aims likely require more promotion, especially in the Significance section. This could be a communication issue. Applicants/investigators should not be afraid to rethink the scope of an application and consider not submitting an aim on resubmission, in particular if that aim is in an "exploratory" or "additional knowledge" category compared with the stronger aims.

LEARNING TO INTERPRET THE SCORES

Exhibit 8.1 shows, under the current review process, a potential means of plotting the criterion scores and the Impact Scores so that the patterns of discrepancy among reviewers, and the gradient of the scores relative to the quality of the aims and the feasibility of achieving them, can be quickly ascertained. The exhibit is easy to develop and provides a simple numerical and spatial snapshot of the scoring and the areas that may need more attention.

HOW CRITERIA SCORING CONTRIBUTES TO THE OVERALL IMPACT SCORE

An article published in 2016 by Lindner et al. provides excellent insight into how the scores for the separate criteria relate to the overall Impact Score. Their data represent 19,719 reviews of new and competing continuation R01 applications. The numbers of initial and resubmitted applications were not provided; nevertheless, they gathered a good sample of scores from meetings held between May and July 2013. The authors pointed out that this period is about 4 years after reviews were modified to include five criteria scores and an overall Impact Score. Figure 1, on p. 242 of their article, depicts the distribution of the criteria scores, which shows a significant skew toward the scores of 1 and 2 for Environment and Investigators (skewness = 1.917 and 1.400, respectively). Innovation and Significance still show a skew toward lower end scores, but the skew emphasizes scores of 2 and 3, respectively (skewness = 0.893 and 0.872, respectively). This figure shows that the Approach Score

EXHIBIT 8.1. Score Agreement Disparity Graph

Impact Scores (Date)
Study Section and SRO (name)
Program Officer (name)
(email address; phone #)

SCORES	1	2	3	4	5	6	7	8	9

AIMS commentary from Significance Section & Innovation
Reviewer 1
Reviewer 2
Reviewer 3
Reviewer 4
Panel Input

SIGNIFICANCE

Reviewer 1	1	2	3	4	5	6	7	8	9
Reviewer 2	1	2	3	4	5	6	7	8	9
Reviewer 3	1	2	3	4	5	6	7	8	9
Reviewer 4	1	2	3	4	5	6	7	8	9

INVESTIGATORS

Reviewer 1	1	2	3	4	5	6	7	8	9
Reviewer 2	1	2	3	4	5	6	7	8	9
Reviewer 3	1	2	3	4	5	6	7	8	9
Reviewer 4	1	2	3	4	5	6	7	8	9

INNOVATION

Reviewer 1	1	2	3	4	5	6	7	8	9
Reviewer 2	1	2	3	4	5	6	7	8	9
Reviewer 3	1	2	3	4	5	6	7	8	9
Reviewer 4	1	2	3	4	5	6	7	8	9

APPROACH

Reviewer 1	1	2	3	4	5	6	7	8	9
Reviewer 2	1	2	3	4	5	6	7	8	9
Reviewer 3	1	2	3	4	5	6	7	8	9
Reviewer 4	1	2	3	4	5	6	7	8	9

ENVIRONMENT

Reviewer 1	1	2	3	4	5	6	7	8	9
Reviewer 2	1	2	3	4	5	6	7	8	9
Reviewer 3	1	2	3	4	5	6	7	8	9
Reviewer 4	1	2	3	4	5	6	7	8	9

PANEL IMPACT SCORE
Cohesiveness of Comments: Strong ( ) Moderate ( ) Weak ( )

moves to a more normal distribution, with the scores distributed around 4 and 5 as the peak with a slope on either side (skewness = 0.243). The mean scores for Approach and Overall Impact are slightly over 4. The means for Environment, Investigators, Innovation, and Significance are roughly 2, 2+, 3, and 3, respectively. The correlation between the Approach Score and the Impact Score is calculated to have a partial $R^2 = .72$ when all five criteria are entered into a regression equation. Approach was entered first (a univariate correlation of $r = .85$ with Impact Score), and Significance as a second predictor accounted for 4% additional variance. Innovation, Investigators, and Environment accounted for less than 1% of the remaining variance. The authors warned that the collinearity is high; the range of scores is restricted; and, when the variables are entered in the reverse order, or varied order, the influence of the individual criteria evens out, with Approach accounting for, at best, 21% of the variance in the Impact Scores.

In essence, the data from Lindner et al.'s (2016) article confirm that the criterion score for Approach tends to dominate in the selection of an overall Impact Score. The data they provided and analyzed verify what most experienced reviewers, SROs, and POs would expect. It is hard to score well in a Study Section when the methodology is receiving most of the negative comments.

MANAGING THE DISTRIBUTED-COGNITION PROCESS

Once the reviews have been carefully examined and the concerns initially catalogued, there are other issues to consider, such as the ability of the applicants to reduce negative commentary in the Summary Statement by examining presubmission activities. Exhibit 8.2 shows a list of presubmission questions that can be used to replicate or change future grant production procedures.

WHAT IS THE PROBABILITY OF RECEIVING FUNDING?

The NIH Data Book shows the potential for success if applicants stick with it. Its website (https://report.nih.gov/nihdatabook/category/6) indicates there are roughly 78,000 applications processed each year for the 21 institutes and six funding centers at NIH. The overall success rate is about 18% for Research Project Grants (RPGs), with *success rate* defined as success per grant, not success per submission. Some mechanisms do not allow multiple submissions, but most of the RPGs do. The success rate of applications based on previously funded grants is close to 50%. A review of the NIH Data Book site shows that success rates differ by institutes and centers as well as by mechanism and

EXHIBIT 8.2. Postsubmission Self, Team, and Laboratory Evaluation

Topics that can help guide questions when evaluating whether and how to replicate or change future grant production procedures:

- Review of the literature
- Data collection
- Data analysis
- Quality of lab or team meetings
- Supportive data publication
- Supportive paper presentations
- Team development, communication, and cohesion
- Internal administrative negotiations and compliance
- Policy comprehension
- Time management and writing of the application
- Compliance to format and policy
- Budget development

Activity Code. The site also shows that some NIH institutes have more funds than others, but that can be offset if there are more applicants to fund. Some of the more well-funded NIH institutes have some of the lower success rates. For many NIH applications, there is no end to the number of times an application can be resubmitted, although the policy for submission changes after two submissions and, of course, the applicants must continue to have the resources to resubmit. The potential for funding from the other NIH institutes claiming an application is not provided in the NIH Data Book. For applications that are close to the funding line, working with the other institutes that have claimed an application is worth discussing with the primary institute's PO. Sometimes institutes are eager to share funding.

THE SUMMARY STATEMENT PROVIDES NEW KNOWLEDGE

How the knowledge generated by the Summary Statement is used after the review tells one as much about the review system and the reviewers as it does about the applicants and the proposed science. If the feedback is crisp and on target with respect to strengths and weaknesses, the system works. A clear set of reviews and a cogent Summary Statement is as important to the PO as it is to the applicants.

Primary reviewers have access to the previous Summary Statement when reviewing a resubmitted application, and they may be the same reviewers who evaluated the original application. Renewal applications include the Summary Statements from the initial funded application. Reviewers can guess the impact of their statements on the applicant's/investigator's funding, but the review process is designed to limit that influence and urges reviewers to just comment on the science. Programs do the funding.

When selecting a book to read, it has been suggested that the book often chooses the reader (Young, 2019). In the grant review process, the content of the application and the structure of the review section narrow the selection of the reviewers by the SRO. In a sense, the topic of the application chooses the reviewers. The goodness-of-fit depends on both the reviewer and the applicant. If a particular application does not seem to have not a good fit with a Study Section, it is worthwhile to think about why this is the case and discuss it with the PO, who might suggest a discussion with the SRO. Applications proposing new approaches or pursuing theoretical issues are particularly important to promote to the SRO or the PO before submission to see whether the goodness-of-fit between applicant and reviewer can be improved.

IN-PERSON REVIEWS VERSUS VIRTUAL REVIEWS

In the December 3, 2020, version of the *Open Mike* blog, Bruce Reed provided a report dedicated to providing several measures that compared in-person versus Zoom meetings. Responses regarding reviewers' approval of Zoom are mixed, with some aspects of conducting reviews via Zoom appreciated (travel issues are reduced, in particular for those coming from the West coast or those who must find sitters for children and/or pets), but those who have traditionally done reviews in this way miss the in-person contact with colleagues and the SRO. There is an organic component to in-person peer reviews that is lost with Zoom. Meeting and noting the expressive mannerisms of panel members via Zoom is a bit like trying to understand a neighborhood by exploring it on Google Maps. The outline of what happens in a Scientific Review meeting will be there, but the effortful organic feeling of traveling to a meeting, attending the meeting, watching colleagues work to express and defend viewpoints, attending a dinner, and traveling back home after the meeting are lost. It is not clear what the change in the distribution of cognition will bring to the concept or construct of a Study Section.

STRUCTURAL RACISM

NIH recently raised the issue of potential structural racism in the review process that apparently has resulted in Black applicants receiving (by statistical examination) fewer funded applications than half those of White applicants (Hoppe et al., 2019; Lauer, 2019). An analysis that summed data over multiple Study Sections (157,549 R01 applications, both new and renewals), from fiscal years

2011 through 2015, revealed three differences in outcomes between Black and non-Black applicants: (a) Blacks had greater interest in topic areas that receive less funding, (b) Blacks were more likely to appear in Pathway 2 (ND) than Pathway 1, and (c) Blacks had higher (less competitive) Impact Scores for applications discussed in Pathway 1. The NIH report focused on outcomes, but the components of the grant review process that contribute to such differences will require a more in-depth examination of the distributed nature of the sub-mission/review/funding process. Where in all the components of the process does something such as systematic bias find its way into the pathways that lead to preliminary and then final scoring of applications?

LIMITING FUNDING FOR MORE ESTABLISHED INVESTIGATORS

On May 2, 2017, NIH addressed funding issues for new investigators by pro-posing a limit on the number of grants a Principal Investigator could receive based on a point system assigned to each Activity Code (https://www.niddk.nih.gov/research-funding/process/apply/new-early-stage-investigators). Feedback from the established research community and the sheer difficulties of main-taining a point system in such a fluid environment as grant funding resulted in a change of plans such that the emphasis would be on providing funds to new investigators by establishing a different review process (clustered by status) for early and new investigators and a more advantageous funding line.

PROPOSED CHANGES IN CSR APPLICATION SCORING

There currently are changes proposed in the NIH CSR review processes that are designed to reduce bias in reviews, improve the integrity of the review process, reduce the burden on reviewers, increase the staffing for CSR, and subsequently increase efficiency and reduce individual burden on CSR staff. CSR has proposed the development of a strategic plan over the years 2022 through 2027. Part of that a plan is to update and simplify review criteria (see Byrnes & Lauer, 2022).

The proposed changes to the scoring process include rolling the elements of Significance and Innovation into a single factor and developing three factors of scoring that will focus on three aspects of the proposed research:

1. Should it be done?
2. Can it be done well?
3. Will it be done?

The areas of emphasis correspond to the three proposed areas of evaluation:

- Factor 1 (officially referred to as *Importance of the Research*), which covers the issue of whether the research should be done. The currently scored components of Significance and Innovation will be combined on the current scale of 1 to 9, with 1 as the best score.

- Factor 2 (officially referred to as *Rigor and Feasibility*), which addresses the issue of whether the research can it be done. This essentially is the Approach category, and it will be scored from 1 to 9, with 1 as the best score.

- Factor 3 (officially referred to as *Expertise and Resources*), which covers the issue of whether the research will be done and whether it will include the previously scored areas of Investigators and Environment. NIH proposes that this factor should be scored not as 1 to 9 but instead be evaluated with a drop-down menu that includes the options of appropriate or additional resources needed. Written justifications must be provided if additional resources are needed.

The purpose of changing the scoring for Environment and Investigator is to reduce the influence of the Environment when considering whether the research will be done. The potential for reviewers to suggest that an investigator is not qualified will likely be reduced as well, and this should reduce the potential for reviewers to express personal bias that can result in a request for a re-review.

Consideration of all three factors will contribute to the final Impact Score, which is still evaluated on a score of 1 to 9, with 1 as the best score. Given that all three factors will include the previous review criteria and will be considered for scoring, it will be interesting to see whether these proposed changes, if implemented, result in funding being spread over a larger number of supporting institutions. The NIH institutes (e.g., National Institute of Mental Health [NIMH], Cancer Institute, NIA, National Institute of Allergy and Infectious Diseases [NIAID]) and other NIH funding sources (e.g., Office of the Director, Office of Research for Women's Health) decide how to allocate resources to Activity Codes and mechanisms as the budget relates to their strategic plans. These individual funding source decisions will have to be taken into account when evaluating the total effect of altered review criteria designed to spread the funding among System 2 institutions. It is possible that the scoring changes will have differential effects with different levels of funding.

Data provided on the NIH RePORTER website (https://reporter.nih.gov/) and research conducted by Eblen et al. (2016; discussed in Chapter 12, this volume) show that current methods, which require scores for the Significance,

Innovation, Approach, Investigator, and Environment sections, results in the greatest range of scores for the Significance and Approach categories. Significance and Approach are the two categories that have the highest correlations with the final Impact Score. The greater variability in reviewers' scores for these categories likely contributes to the potential for higher correlations, but it makes sense that Significance (including the occasionally hard-to-define Innovation category) and Approach would be expected to have the greatest influence on the final Impact Score.

The proposed scoring method (e.g., Factors 1, 2, and 3) emphasizes the scores (Factor 1 and Factor 2) that historically provide the greatest value relative to the final Impact Scores while reducing the impact of the applicants'/investigators' System 2 reputation to Acceptable or Not Acceptable. If submitting from a more frequently funded System 2 institution reflects greater resources in that System 2 institution, then not much will change in the probability of funding for System 1 applicants/investigators.

The CSR application also makes an effort to distinguish between Additional Review *Criteria* and Additional Review *Considerations*. These categories continue to add to reviewers' burden, and CSR is proposing to reduce the onus of Additional Review Considerations by requiring that Data Management Plans now mandate that data sharing be reviewed by the appropriate NIH officials; otherwise, there are few changes in the requirements for reviewers and applicants in regard to the need to consider the Additional Review Criteria and Additional Review Considerations.

The Additional Review Criteria are not scored but can be considered when providing a final Impact Score. The proposed instructions list the Additional Review Criteria as follows:

- human subjects protections

- inclusion of women, racial and minorities, and individuals across the life span

- vertebrate animal protections

- biohazards

- commenting on resubmission quality when an application is resubmitted and responds to the previous reviews

- commenting on the productivity of renewals: Was the previous grant productive, and does it promote the aims of the renewal application?

- commenting on Revision applications where revisions can be small budget additional studies or large budget Competing Revisions requiring a full review (Byrnes & Lauer, 2022)

These categories are proposed to be scored as Appropriate or Concerns. A rating of Concerns requires a narrative to explain those concerns.

The proposed Additional Review Considerations include

- authentication of key biological and/or chemical resources (evaluated as Appropriate or Concerns) and

- budget and period of support (rated as Appropriate, Excessive, or Inadequate).

At face value, none of the proposed changes in peer review should affect the writing of the aims or the other sections of the typical grant application, but subtle changes could develop.

In addition to proposed changes in peer review, CSR has laid out an ambitious plan, including the use of artificial intelligence to more accurately and efficiently refer applications to Scientific Review Groups (SRGs) and specific Study Sections. This proposed change would result in the tailoring of documents to the Study Section's preferences and could be as influential on scores as changing the review criteria. NIH and CSR have access to significant amounts of data and make decisions and produce reports on the basis of those data. Applicants/investigators should keep in mind that the cycles of review and funding make today's data reports a reflection of at least 2 years of functioning; that is, program changes can be monitored in real time, but it takes a few years for data to suggest that a change in process has resulted in a trend in reviewer evaluations and subsequent funding by NIH institutes, centers, and offices.

KEY TAKEAWAYS FROM CHAPTER 8

- The review process produces a product called the *Summary Statement*. This statement results from the need to transmit applicant/investigator-generated research plans to the reviewers for evaluation and then transmit the evaluation back to the applicants/investigators. Communication between applicants/investigators and reviewers is not done one to one but is distributed and shaped by electronic submission formats and the constraints that submission and response formats may place on the communication of ideas and evaluation of ideas.

- Each funding agency or source has its own submission and review process that is communicated to applicants/investigators in a FOA/NOFO. There are similarities between all review processes that ask applicants/investigators to write as form of communication and explain (a) the goals of a project;

(b) support for the premises of those goals; and (c) if the project is a data-generating one, a plan to carry out the goals and analyze data. The degree of elaboration of feedback in the form of a Summary Statement is related to the resources of the funding source.

- For some funding agencies, peer review and funding are separate processes (e.g., NIH), and for other agencies the review process and initial recommendation for funding are combined (e.g., NSF) at the initial stage of review and followed by a second level of review for funding.

- Funding agencies that receive large numbers of applications, such as NIH, may introduce a triage process into the review process whereby only the upper half (best) of the applications are discussed in peer review and receive, in the Summary Statement, a report on that peer review discussion. The triage process is based on preliminary scoring by assigned reviewers and rules for assigning applications to triage. The triage process has been criticized, but it is necessary to manage the reviewers' time and the funding source's finances.

- The funding process that follows a review can take several months, during which applicants/investigators must consider whether they should prepare for a resubmission if funding does not seem imminent. For NIH, a triaged application indicates that a resubmission is required, but otherwise NIH applicants must rely on published funding-line information that will indicate what scores are fundable.

- One of the most important skills applicants/investigators can develop is understanding how to respond to concerns reviewers note. Reading reviews that are not favorable can generate anger and anxiety. Several suggestions were made in this chapter regarding how to focus on positive feedback as well as negative feedback. Many applications are not funded on the initial submission, and learning how to read reviews and resubmit an application is a part of becoming grant literate.

- Funding agencies can change their criteria for review, and NIH has recommended changes in the review process for some Activity Codes that will reduce the influence System 3 reputation has on the peer review process.

9

TRAINING GRANTS

As the result of being exposed to a STEM education (Discovery Today, https://www.nsta.org/), many individuals in high school and undergraduate school believe they would like to pursue more education in STEM areas and possibly become involved in research. Making the decision to pursue further education in an area of science requires that one consider the potential for

- understanding the multiple educational pathways toward a career in science,
- successfully completing the next level of training,
- maintaining self-interest and satisfaction,
- regaining sunk costs of time and money via future employment and employment satisfaction, and
- flexibility of use of training.

There is nothing unusual about wanting to make good choices, which will continue if there are potential choice points in one's education and career development.

Figure 9.1 shows the typical upward steps of education that support a career in science. The arrows in the figure represent transitional shifts in one's

https://doi.org/10.1037/0000390-010
Get Funded: A Practical Guide to Understanding the Grant Application Process and Writing Winning Proposals in the Behavioral and Biomedical Fields, by J. W. Elias

FIGURE 9.1. Upward Steps of Education and Inflection/Transition Points That Lead to a Career in Science

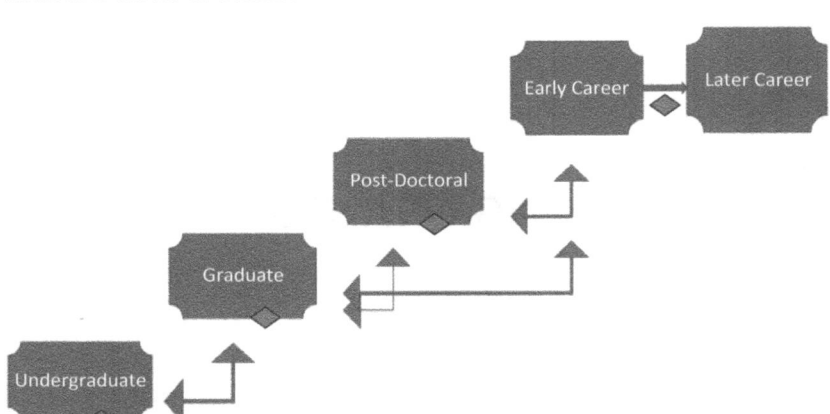

personal education, such as from an undergraduate school to a graduate school, to a postdoctoral position, to a potential permanent position. If there is the potential for more training to provide predictable employment, and perhaps higher wages or more personal interest, one likely will have more interest in pursuing further training. The two-way arrow connections represent the strong need to establish the previous level of training as a base for the next level of training. The diamond shapes in the figure represent likely inflection points at which appraisals are made of one's current level of skills and recommitment or a shift in commitment. Each career period represents about 4 to 6 years, indicating that an early science-as-a career stage could be reached in a minimum of, say, 8 years, including undergraduate and graduate education, but more likely the move from an undergraduate education to an early career stage requires 10 to 11 years. For careers that involve clinical training, the transitional periods may be longer, and there will be additional inflection points within the clinical training experience.

Doubt, hope, and building positive expectations are emotional components of the thinking-through process at all levels of training and career development. In more predictable environments, there are inflection points at which one must take the time to take stock before moving forward. For example, in undergraduate school a typical transition to a declared major in a science-related field often occurs in the sophomore year or as late as the junior year. As intellectual interest in science grows, in the best of all worlds there should be a developing interest in the activities and basic skills of science as well.

Transitions from one level of training to another are made more easily when the basic skills at the current level of training can be used to obtain the new skills needed at the next level. For example, within-laboratory postdoctoral environments can be competitive. Individuals hoping to use a postdoctoral position to acquire specific research skills (e.g., laboratory skills, statistical analytic skills) should check with their potential sponsor to learn whether those skills should already be in place or whether they can be obtained without slowing postdoctoral progress. Some laboratories that seek to maximize the potential for success in their new fellows provide a navigator who is assigned to guide the new addition to the laboratory through the new environment and the new set of expectations.

For people who have committed to a career in science, each step in their career direction requires another degree of commitment, goodness-of-fit, and a matching of their expectations of themselves and others. In graduate school, somewhere between the master's and the doctoral degree, there is often a re-dedication to a career in research or a career that is research related. Of course, financial debt and the potential for a job after training are two of the determinants of goodness-of-fit with the potential next education/career stage.

Looking ahead at the requirements and eligibility for training awards can reveal where the windows of opportunity are most likely to be available and leveraged. The specifics of the awards, and your eligibility for the awards, will change over time, but if you are contemplating a career related to research it is always advantageous to look for advancement opportunities that can be leveraged early in the education process (e.g., a summer devoted to working in a laboratory to acquire a new skill). As explained in Chapter 1, part of acquiring grant literacy is not just what you remember but what you pay attention to. Paying attention to opportunities is very important to the advancement of knowledge in a science-related career. For those interested in a clinical career, researching the timelines for submission and completion of career awards may be time consuming, but funding institutions can make accommodations to fit the extended time needed to acquire practical clinical training.

Career-advancing awards at funding agencies (e.g., National Institutes of Health [NIH]) often refer to a specific kind of award funded under a particular award Activity Code (e.g., NIH fellowship awards F31, F32, or NIH career awards K01, K08). NIH refers to *K awards* as advanced career opportunities that require greater experience and advancement compared with fellowship awards. Individual fellowship awards are funded as Ruth L. Kirschstein National Research Service Awards using the F activity code. Fellowship training is also provided by institutional (T) training awards. These terms, and others, can be found in the NIH glossary (https://grants.nih.gov/grants/glossary.htm).

The National Science Foundation (NSF) refers to its advanced career awards as CAREER awards. The Department of Defense (DoD/DOD) offers Young Investigator awards and Mid-Career awards.

A BASIC UNDERSTANDING OF CAREER DEVELOPMENT AWARDS

Multiple federal agencies, including NIH, NSF, and DoD/DOD, offer a number of grant opportunities to support research at each level of career development. In addition to resources available from those federal funding agencies, most colleges and universities have Offices of Sponsored Programs that can direct applicants to appropriate financial resources for education and research-specific financial support. This would include potential financial support from professional societies and associations as well as from private funders.

General information on the fellowship and career awards and FAQs can be found on the funding agency websites (e.g., NIH F Kiosk and K Kiosk; https://researchtraining.nih.gov/programs/fellowships and https://researchtraining.nih.gov/programs/career-development, respectively). The NIH career training website (https://researchtraining.nih.gov/programs/fellowships) and the fellowship websites at NSF provide excellent introductions to the available programs. NSF has a graduate fellowship program (see https://beta.nsf.gov/funding/opportunities/sbe-postdoctoral-research-fellowships-sprf), and DoD/DOD offers peer reviewed career development awards (see https://research.njit.edu/dod-peer-reviewed-career-development-award). These sites clarify the general timelines of training required for each award and the eligibility requirements. Information on salaries and benefits can be acquired by searching online. Salaries and benefits can change or increase from time to time, so applicants/investigators should verify that they are working with the most recent information. In addition to the websites that provide more general information, applicants/investigators should locate the websites for the F and K awards for each NIH institute or NSF program to see whether the specific institute or specific program/directorate information differs from the more general information announcements. NIH provides a document titled *Review Criteria at a Glance* (https://grants.nih.gov/grants/peer/guidelines_general/Review_Criteria_at_a_glance.pdf) that identifies the review criteria that have been established for fellowship, career, and institutional training. At NSF, applicants/investigators can look for information that might be specific to the eight directorates. NSF and NIH conduct reviews of training applications in a similar manner. When contacting a Program Officer (PO) to discuss a research project for a fellowship (F award) or an advanced career award (K award), it is a good idea to ask where in the review process the pressure points are.

POs at NIH will be adjusting to the new individual fellowship review criteria proposed by the Center for Scientific Review (CSR), which manages reviews for the majority of individual fellowship applications. The new criteria were discussed April 25, 2023, by CSR Director Noni Byrnes and NIH Deputy Director for Extramural Research Mike Lauer in the *CSR Review Matters* blog, which cross-references the NIH *Extramural Nexus* blog. Both blogs can be subscribed to and are likely the best outlets for keeping up with changes to review and funding policies. The proposed changes for the review of NIH fellowships are discussed at the end of this chapter.

In general, training in research methods in some form is best accomplished early in one's education. Seeking more training and committing to a specific path of training and education are parts of research-oriented career development. A significant component of research training is understanding ethical standards and the responsibility of researchers and their supporting organizations. Education in the responsible conduct of research (RCR) should be acquired in some form at all levels of research training. NIH and NSF require RCR training for anyone supported on a grant (see https://ori.hhs.gov/general-resources).

Undergraduate and graduate students can gain access to research resources via mentors who are themselves supported by institutional resources available through doctoral programs; training awards (T awards); or research program awards, such as R01s. Most research training at the graduate level is supported by System 1 or System 2 means other than training grants or fellowships. NIH Data Book Report 235 (https://report.nih.gov/nihdatabook/report/235) illustrates how the means of support for graduate students remained very steady from 1985 to 2020. Graduate students often receive research support by working as an unpaid research assistant in a professor's laboratory but receive course credit and anticipate taking part in a paper presentation or receiving publication credit. They may be financially supported by a research grant that has been awarded to a mentor or by a T award awarded to an institution for the purposes of group training and supporting graduate students.

Postdoctoral positions are almost always funded positions. Most funding comes from research grants awarded to mentors but, as noted, individual research fellowships can be applied for and provide a degree of prestige and independence, although they are still mentored. It is important to note that, to be competitive, applicants for individual F and K awards typically require mentors who are grant funded. Funding for career awards is awarded to the System 2 institution for the support of the applicant and the project.

In some cases, postdoctoral students may work in a specific laboratory but are supported by an institutional T award to a group of researchers who are focused on an area of research.

There is no evidence to verify that support from a federal institutional training grant or an individual fellowship grant provides any better training than that which is supported by a graduate program and credit hours or by a mentor's research grant. The funded training programs may provide additional resources, such as travel, to students. Students supported by federally funded training programs should always ask what the extra benefits are.

The benefits of training within a research environment depend on the quality of the research experience and how well that experience fits the trainee's interests and goals. Nevertheless, there is a return-on-investment component to having applied for and received a fellowship award or career award, in particular as these awards relate to future employment or grant funding. However, grantees must complete the proposed project to provide a return on investment to maintain the reputation of the grantee, supporting mentor, and supporting institution. Therefore, when postdoctoral students decide to seek financial support by pursuing an individual fellowship award it is extremely important to be able to work within a budget and a timeline that will allow the project to be completed. There is no guarantee that System 1 mentors or System 2 resources can be applied to an incomplete federally funded grant if fellowship budgets are exceeded before the end of a proposed project's end date.

The process of working with a mentor, or mentors, to develop a fellowship application is an excellent introduction to the world of grant funding. Applying for a funded fellowship status not only indicates a level of vetting by mentors but also signals a move by the applicant toward independence in research and research support. There are only a few published studies based on NIH data that have investigated the future benefits of a fellowship award aside from financial support during the fellowship period. Conte et al. (2020) examined NIH fellowship data from 1996 to 2008 and found a positive relationship between fellowship status and future funding success. Much of that relationship is likely due to the greater probability of fellowship applicants being supported in universities that are in the top tier of funding (Pickett, 2019).

Pickett (2019) found that, below the top four institutions, there was no clear relationship between institution of training and specific number of research dollars later obtained by fellowship and career awardees. Pickett's findings are better understood if one has a good knowledge of NIH function and the nature of funding, but one thing that is clear from the article is how difficult it is to pin down the future benefits of a specific award when a funding institution like NIH has so many individual institutes and so many versions of career awards that come onto the scene. Several years are required for the awards to be completed, and then there are several years that follow the career awards when other grants, such as Research Project Grants (RPGs), are submitted before they are eventually funded. Therefore, the effects of research and career awards are often mediated through the opportunities they make available relative to other kinds of career development support.

The personal timing for applying for a fellowship award must fit with the guidelines of the application. Becoming familiar with fellowship opportunities (mechanisms and announcements) can help with planning for the personal opportunity to apply. Fellowship applicants should not interpret the nonfunding of an application as a portent of things to come but should instead use the experience of applying as a window into the competition for research funding. Fellowships are highly competitive, and planning with the help of mentors can make an applicant more competitive.

Reviewed but not funded predoctoral fellowship applications can be used to develop a postdoctoral application that in the future can be used to develop an R01 grant application. The experience of the submission process itself can be self-diagnostic. If the process of applying seems too intense, then following a career path that involves grant funding for support or promotion might not be a good choice.

THE APPLICANT IS THE FUTURE

A model that can help direct the writing of any training grant mechanism, the "Applicant Is the Future" model, is presented in Figure 9.2. This model envisions the fellowship applicant or career candidate (A) as a raw product who works with the mentor (B) in a support environment (C) to provide the reviewers with a vision of how the training process (D) that will produce a more competitive scientist who can compete in the current and the future research market. This is a transformative model in which the raw product (A) is transformed via training into a final product of an experimenter with multiple dimensions (E+).

The final human capital product of the A-to-E+ process should be a scientist who is capable, by means of interest, training, and experience, to continue

FIGURE 9.2. The "Applicant Is the Future" Model

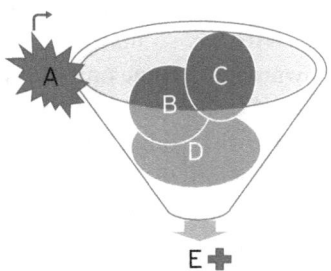

Note. A = the raw product (applicant); B = mentor or mentors; C = the support environment; D = the training process; E+ = an experimenter with multiple dimensions.

in research. All applicants, but in particular applicants at the K award level, should ask themselves how much more capable they will be after completing this process and whether there are other pathways that could lead to the same result. The crystal ball component of career applications requires that one estimate whether training will enhance or limit employment opportunities. At the fellowship level, applicants should keep an eye on the employment options to which extra training will lead. A resources-heavy specialty (i.e., one that requires access to technology or colleagues with specialized training) can limit the future System 2 employment options while increasing the personal resources of the System 1 investigator. The clear implication from most funding agencies is that a training experience will result in more future grant applications from the trainee, which will lead to more support for other scientists and more publications.

To further illustrate the model in this schema of financial support for training, the System 1/2 (B) mentor (or mentors) are the production supervisors and managers who direct and partially finance product development. Thus, the applications for F awards and K awards place a focus on mentor involvement beyond simply the availability of a mentor. The mentor's laboratory (or laboratories) is a part of the larger (C) System 2 institutional environment that could be considered the factory for product development—therefore, the applications require demonstration of previous and current System 2 support for training and research production. The research proposal/training (D) is part of product development and extensively involves proposed interactions between the applicant/candidate and the mentor as well as any classroom and laboratory experiences that would support and promote the skills needed to move to the next level of research expertise. The D component is the integrated mentor-in-action/curriculum/research project element, the action portion of the proposal that must describe how all the parts will fit together. A smooth integration of B, C, and D is important because many promising applications do not make it to funding status because of a clear lack of integration of the research project with training in the D component. It is the combination of B, C, and D that results in the transformation of A into the final product, E+. CSR's proposed changes to fellowship reviews will influence component C of the "Applicant Is the Future" model more than other components; that is, System 2's reputation and mentor achievement will be deemphasized for reviewers. This change is discussed further at the end of this chapter.

In the case of the System 1 fellowship awards, the goal is to move toward independence as a researcher while still receiving guided experience from an experienced mentor. Postdoctoral applications can show how one's previous dissertation research helped develop the postdoctoral plan, but the postdoctoral application should not be simply an embellished version of the dissertation. Nevertheless, a forward thinker could shape a master's project to support the premise of a dissertation, and both could provide the pilot data for an independent fellowship application as opposed to a System 2 institution-supported fellowship. In the case of the K series career awards, the goal is to access training and research resources via mentors and the sponsoring System 2 institution to develop even further in the chosen field as a potential System 1 independent leader in research who will continue working in the sponsoring System 2 environment.

The need for the System 2 sponsoring institution to clearly indicate current support status and future support status for applicants, even those who do not receive a career award, is very important. Sponsoring institutions that can show that there is solid ground for career development are more competitive in the pursuit of career development awards. Large research institutions and medical schools have a number of employment status designations that are not well defined to those outside of the institutions. A means of moving from a lower level position to a higher level position in a System 2 institution would be to write a successful career development award that could result in a higher status that will help support further grant applications and lead to more independence. One's postaward position in the supporting institution must be made clear. Lukewarm letters of support sent to System 3 review panels from System 2 make a System 1 applicant less competitive.

Data from a 2018 National Institute of Allergy and Infectious Diseases (NIAID) analysis (https://www.niaid.nih.gov/grants-contracts/research-career-development-k-award-success-rates-future-award-prospects) support the contention that recipients of K awards have an advantage in applying for and receiving further funding. The investment component for the applicant

and the institution is a positive one in general. Everyone has their own context in which a training award would be beneficial.

SYSTEM 2 INSTITUTIONAL TRAINING

T awards work within the same conceptual model as the "Applicant Is the
Future Model," except the proposed class of applicants/candidates as whole
are component A in Figure 9.2, and recruitment and retention of this component are the focus of preaward evaluation and postaward application.
Component B for institutional training grants is a cadre of skilled mentors
who will provide support for C, a circumscribed area of training, to include
curriculum and research. Component D is the strength of the support institution, including the brief history of other training programs cosponsored
and colocated at the institution and the proposed interaction of the new
training program with the training program that makes up component D.
Part E+ is the potential for the recruited appointees to make it through the
program and develop as promised.

The development of training awards requires a substantial promotion of
the training institution, the mentors, and the means of recruitment and retention of applicants. A successful application for an institutional training award
is just the first step toward showing that the program can work as advertised
and continue as a training program. The renewal award is the real stamp
of approval for a training award and, by association, the individual fellows.
These are record-development– and record-keeping–intensive awards. Institutional training grants attract applicants to a specific area of training and
offer mentor-guided experiences and research support. The better institutional training applications promote specific skill training as well as conceptual training. It never hurts for a System 2 T award to produce graduates who
go on to research careers and use their training in research.

Research assistantships also provide a pathway to training as a researcher,
and many long-term grants, including program projects and center grants, can
be prolific sources of training and career development funding. The competition for System 2 institutional training grants is often more local than other
means of support, but it might not be, especially at the graduate level.

NIH and most institutions that support research training or research as a
career have a focus on supporting racial and ethnic minority applicants or
attracting diverse candidates. For funded researchers with a couple of years
remaining on their grant, the submission of a competitive supplement to support
a project related to the aims of a grant but that offers a new component is
a common method of support and is often used to attract diverse applicants to

work with ongoing grants. A number of Funding Opportunity Announcements/ Notices of Funding Opportunities (FOAs/NOFOs) specific to NIH institutes are offered to enhance this process. Diversity can include in its definition financial inequity or physical disability.

For applicants who committed themselves to research-oriented careers as students, there is an implicit commitment to competing for research-focused entry-level jobs in academic or independent research institutions and businesses. The better competitors have more experience; a better match with employers' needs; and evidence of research capability, including publications and signs of independence. Participation in fellowships and institutional training grants typically provides a competitive edge for employment because it illustrates a form of preemployment vetting and commitment to processes of training that other research experiences may provide, but not as obviously or competitively.

The choice to compete for a fellowship award gives the applicant the opportunity to be introduced into the federal funding process with guidance from experienced mentors and the potential to be recognized early in a research career for a move toward guided independence. There are time and effort commitments for all involved (applicants, mentors, support personnel at the submitting institution). If applications are not funded on the first submission, there is often the opportunity to revise and resubmit if this process fits the time frame of the applicant and the mentors. As noted, fellowships are not a prerequisite to a research career, but they do provide experience in submitting and receiving feedback for a proposal that is typically competing on a national level. If the application is fundable, then applicants must make the decision of whether to accept and commit to a pathway of training or choose another pathway.

Applying for and receiving a fellowship offer, but accepting a more attractive offer instead (e.g., the choice between a successful postdoctoral 2-year award and a tenure-track university position), is acceptable behavior. Accepting and starting a fellowship and then wandering off track, on the other hand, can be a risky decision. Leaving an award also affects the mentors. This goes back to understanding the degree of commitment associated with grant funding and the context of goodness-of-fit to career goals and between the expectations of the applicant and mentor.

All applicants for training awards must present themselves as they are now, including their story of educational and training development (A in Figure 9.2). Just as providing a personal story of research interest and involvement is a unique component of training awards, so too is using this history as a baseline for describing what training will produce (E+). The initial written component of a fellowship or career award application, as well as the first evaluative component of the review process, is to review the personal story of the applicant as it relates to early interest in research and accomplishments.

Accomplishments could include significant experiences, paper presentations, publications, and awards. Recall the inflection and transition points as depicted in Figure 9.1. This figure keeps the timeline intact and helps explain the motivation for moving toward the E+ designation. Individual fellowship and career award applications are unique in this regard given that most grant mechanisms rely on the applicant's biographical sketch to provide a history of past research and educational experiences, and the inflection and transition points are often inferred.

Returning to Figure 9.1, the applicant's written personal story typically begins at the undergraduate stage and continues to add experiences up to the point of submission. The more compelling stories include an early education interest in research and perhaps research success. Many of the more successful System 2 institutions begin grooming promising students in the early stages of their education for awards at the later stages of their graduate and postgraduate careers. These efforts are somewhat selfless behaviors for grooming institutions because fellowship students do not always begin their immediate career appointments at the supporting training institution.

Keep in mind that the applicant-development story is not told solely from the point of view of the applicant. Part of the story and promotion of the potential to grow as a scientist is told by the supporting letters of reference and the requested comments from mentors regarding training potential. When asking for letters of support, it is important to remind your references of your previous history and your accomplishments. The following list is derived from current CSR suggestions to reviewers. Applicants should be able to provide guidelines for their references and mentors and specify the following:

- research capabilities shown and the potential that the applicant can become an independent researcher,
- whether the scientific and technical experiences of the applicant make sense relative to their goals,
- evidence that the applicant can organize scientific data and provide cogent written and verbal communications,
- quality of research activities and publications that have already been completed,
- whether the applicant has a sufficient knowledge of the literature,
- the degree to which the applicant can stick to a plan and a goal,
- the degree to which the applicant has shown originality, and
- the potential value of the proposed training to the applicant's needs and goals.

EXAMPLES OF HOW TO PREPARE FOR SUBMISSION

When preparing to submit an application for a training or career award, an applicant should first assess their personal commitment issues. Writing a career development application with a mentor is hard work and time consuming for both the applicant and mentor. As it becomes more obvious what is required to be competitive, both the applicant/candidate and mentor should make an appraisal of the following:

- degree of personal commitment to training goals,
- the goodness-of-fit of a task with personal short- and long-term time commitments,
- personal expectations for the task per se to enhance career advancement for all parties involved, and
- whether personal circumstances permit a one-time effort or a stick-with-it-until-success effort.

A personal training award is not just a means of personal financial support or support for a research project; it is a career path application. The successful examples of training awards offered by NIH (https://www.niaid.nih.gov/grants-contracts/sample-applications, https://grants.nih.gov/grants/how-to-apply-application-guide/resources/sample-applications.htm) show how the successful application ties in A (previous experience) with the need for B (mentoring) to implement the (D) integrated mentor/curriculum/research plan. Component C ensures that the supporting institution has the resources and the commitment to support the applicant and the training. Particularly impressive in these examples of funded F and K awards is the commitment of the mentor and their knowledge of the applicant's existing strengths and how those strengths will be directed toward (D) the career development process. Solid research plans can be seen in the examples NIH offers, but it is the integration of mentor, curriculum, and project that is most impressive. In these examples, the degree of commitment; goodness-of-fit among goals, mentor, training, and research project; and the matching of expectations noted in component A with component E+ can be clearly observed.

Often, the inspiration and pursuit of a training award comes from a System 2 administration's goal to have more individual training awards as proof of status. This aspiration may or may not fit with the immediate goals and of the applicant. It would not be unusual for potential career applicants to develop the idea of a research project before thinking of a career development award that will help them accomplish the research goal. The project could involve the interests of several funding institutes and require various skills to accomplish. This situation brings up a goodness-of-fit issue with respect to which funding institute to ask for training and an assessment of how the mentor's professional strengths and interests fit with those of the funding entity. If, for example, the proposed research requires knowledge of aging, infectious diseases, and pathology, to which funding institute should the application be directed, and what should the focus of the training plan be? This is a circumstance in which a research grant proposal with multiple investigators with expertise in each of the primary areas may be a better fit to the initial research goal than a career award that tries to fit multi-investigator research into the confines of a career award.

COMPARE T AWARDS WITH F AWARDS AS A TRAINING SOURCE

NIH undergraduate training grants are available for research training institutions but not for individuals. Institutional training awards are referred to as *T awards*. As noted, T awards are applied for by institutions and are available by specific designations for all levels of training, from baccalaureate to postdoctoral. T applications for individuals are available through the funded institution. Institutional Offices of Sponsored Programs can provide information about the existence of training grants within institutions.

Applicants pursuing a training award should first locate, via the funding institute's website, the most recent Program Announcement (PA) related to the specific award. There is an almost overwhelming amount of information on the web and on NIH websites on how to apply for a fellowship or a career award. Most of it is worthwhile reading, once the basic understanding of a fellowship or career award has been acquired. However, to begin the writing process it is critical to cut through any potential confusion provided by multiple websites by locating the current PA. For example, at the time of this writing, FOA PA-21-051 (https://grants.nih.gov/grants/guide/pa-files/pa-21-051.html) designates the parent (generic) funding announcement for a current F31 announcement. Over time, PA numbers for the same mechanism (e.g., F30, F31, F32) will change because the PA is reissued periodically with new updates.

When reading a PA for a fellowship or a career award, look under "Key Dates" for the posting date of issue and then for the date of expiration. A succession of PA announcements is designated by an indication under "Key Notices" that a PA has been reissued on another date under another PA number. The issue date—for this example, PA-21—was October 21, 2020.

The expiration date for this PA is September 8, 2023. The information in PA 21-051 is viable until then. If there is any addition or update to the information in the PA, it will appear above the PA number in the form of a Related Notice.

LOCATE AND THOROUGHLY EXAMINE THE SUBMISSION FORMS

A PA provides a link to the appropriate forms to apply for a grant mechanism. Applicants should go through the PA and find the link to these forms, which explain and contain the appropriate information for each submission. Printed booklets containing the forms can sometimes be confusing; some of the information is repetitive, with links that can take applicants around in a loop. Nevertheless, the specific and most up-to-date information on application submissions from beginning to end can be found in the current PA, which will have links that lead to the current forms.

In many cases, applicants and mentors are in a rush to get to the task of writing because they have a general sense of what should be in the application or have access to a recent application. *Only the current PA and the current forms will have accurate information.* Reading the PA in detail and examining the submission forms before writing is a reminder of the commitment required to learn how to submit. Many a tear has been shed by applicants and mentors when they attempt, at the last minute, to fit the details of an application onto the required forms, recognize that something does not match, and then frantically search for guidance from the local System 2 administration or the System 3 funding institute.

COORDINATE THE WRITTEN REQUIREMENTS WITH THE STATED REVIEW ELEMENTS FOR EACH AWARD

It is always useful to coordinate the instructions for writing an application with the instructions for the review of an application. Before suggesting changes for the review of NIH fellowships in 2023, the NIH prepared and updated, in 2021, a document that provides a succinct list of the review criteria for Activity Codes, including fellowships, career awards, and training grants (see NIH, 2021). The review criteria for fellowship awards will remain in place until the suggested changes for the review of fellowships become policy. If the changes are made, the policy will likely be in place for 2024 applications.

HOW WOULD CSR'S PROPOSED CHANGES FOR FELLOWSHIP APPLICATIONS CHANGE THE FOCUS OF REVIEW?

The CSR and the NIH *Extramural Nexus* blogs emphasize changing the weight that reviewers place on the reputation of the applicant's supporting System 2 institution and of the sponsor's/mentor's achievements (Byrnes & Lauer, 2023).

The goals for the proposed changes as stated in the blogs are as follows:

The proposed changes will 1) allow peer reviewers to better evaluate the applicant's potential and the quality of the scientific training plan without undue influence of the sponsor's or institution's reputation; and 2) ensure that the information provided in the application is aligned with the restructured criteria and targeted to the fellowship candidate's specific training needs.

The proposed focus for scoring will be on a) scientific potential, fellowship goals, and preparedness of the applicant, b) science and scientific resources, c) training plan and training awareness. From the provided descriptions of the proposed changes the requirement and methods of training will not change, but there will be an emphasis on the description of an applicant's personal preparedness and personal motivation within the strengths or limitations of their past and current training contexts. Achievement is important for the mentors and the applicants, but personal attributes and motivations are important as well. Potentially letters of reference and support will be important with respect to the personal motivations of a candidate.

The three proposed scoring categories (scoring 1–9, 1 is the best score) are:

I. Scientific Potential, Fellowship Goals, and Preparedness of the Applicant
II. Science and Scientific Resources
III. Training Plan and Training Resources (Byrnes & Lauer, 2023)

The preparedness and environment provided for students interested in research at highly funded research institutions will make it hard to move the needle on the numbers of applications coming from System 2 institutions where research is important but the research activity level is more limited, as is the variety of research experiences.

To aid in understanding the review process for fellowship and career awards, two sample fictitious reviews, developed on the basis of my experience and current review criteria, are provided in Exhibits 9.1 and 9.2. Exhibit 9.1 is an example of a review of a fellowship application, and Exhibit 9.2 is an example review of a career award.

Reviewers will provide an overall Impact Score that reflects their assessment of the likelihood that the fellowship will enhance the candidate's potential for, and commitment to, a productive independent scientific research career in a health-related field. An application does not need to be strong in all categories to be judged likely to have a major impact.

EXHIBIT 9.1. Sample Review (Fictitious) for a Fellowship Applicant Receiving an Impact Score of 1

Review Criteria

1. **Fellowship Applicant**

   Strengths:

   - The applicant shows a continual period of achievement and interest in the application of skills in finance/accounting to medical treatment research issues.
   - The proposal is well organized, with materials easily located and understood.
   - Several articles and papers have been produced as well as relationships developed that will support the proposed training.
   - Letters of support are strong and to the point relative to the critical issues in training and support for research.

   Weaknesses:

   - No weaknesses are perceived.

2. **Sponsors, Collaborators, and Consultants**

   Strengths:

   - The sponsors show strong combined expertise and training skills to support the aims of the project and facilitate data analysis and publication.
   - Funding and support for applicants is a strength of the supporting institution and the mentors.
   - The adventurous nature of the research project is supported by the history of the sponsors and recognition that the applicant may need an assigned peer navigator at some of the inflection points of the proposed training and research.

   Weaknesses:

   - Potential weaknesses are offset by recognition of points of inflection in the training process and a plan for addressing any concerns. No overall weaknesses are perceived.

3. **Research Training Plan**

   Strengths:

   - The training goals are well described and are in line with the career history, goals, and self-described projected trajectory for development.
   - The aims are straightforward, progress without overlap, and the data outputs are clear and available.

   Weaknesses:

   - Known potential measurement and interpretation issues are mentioned in a letter of support but not carefully discussed in the proposal. Good to know that the research team is aware but would prefer to have knowledge transferred to the candidate's proposal.

(continues)

EXHIBIT 9.1. Sample Review (Fictitious) for a Fellowship Applicant Receiving an Impact Score of 1 (*Continued*)

- The extension of the brief measurement period over an extended/projected time via modeling could be better vetted relative to the costs and benefits of the measurement process, to include the cost of small intervention sampling.

4. Training Potential

Strengths:

- The training potential is clearly delineated in the clarity of the presentation, including the motivation of the applicant, and the integration of the research and didactic components of the training plan. Given the history of the mentors the likelihood of a positive outcome is very high.

Weaknesses:

- The timeline is ambitious and does not leave much space for delays in progress.

5. Institutional Environment and Commitment to Training

Strengths:

- The research environment and commitment to training are exceptional.

Weaknesses:

- No significant weaknesses are noted, although the potential for a redefinition of the described academic departments and a split across two geographic sites noted in a description of the research environment is a concern. The letter from the department chair emphasizing the completion rate of fellows in the department and the companion letter from the dean emphasizing the need to retain the success rates offsets some concerns about the diverging of the department into two different units.

OVERALL IMPACT/MERIT

The goals of the applicant match well with the training proposal and the skills of the mentors, including the history of funding by the mentors and the provision of success rates for trainees within the primary support department. There are some issues of concern regarding tight timelines and potential disruption relative to geographical relocation of labs. Given the track record of the supporting institution and the department per se these issues are likely to be managed well should they occur. The applicant is matched well with the training program and with the mentors.

Impact Score: 1

EXHIBIT 9.2. Sample Review for a Career Award

An NIH career award should focus on the integration of mentor training with the proposed research project. The proposed project should be of research project grant RO1 quality and lead to an advance in skills, not just data collection.

Candidate

The candidate provides a detailed history of publications ($N = 15$) and research achievements, including a 2-year F32 Fellowship with Dr. Guy Bucky in the Julie Ross Institute at Old Allegheny. The F32 produced two publications focused on verbal working memory loss and aging. The candidate was also awarded the Hugh Gilmour Award for best paper in experimental psychology. Following the fellowship, the candidate took 2 years of family leave before securing a lectureship for 2 years and then for the past 2 years has worked in a staff scientist role at the Texas Technology Institute (TTI), working with one of the current mentoring team members, Phil Marsh (PhD). Dr. Marsh is well known for studies of motor memory; his recent study of motor memory learning and retention assessed pre- and postconcussion high school and college football players. The number of publications by the candidate since arriving at TTI has been steady for the past 5 years ($N = 5$), with two first-author publications. A Small Business Innovation Research grant, focused on a brief screen to assess motivation in adolescents recovering from head injury, was submitted last year with her secondary mentor, Dr. Tree. No outcome for that submission is provided by the candidate. Although the publication record has been strong and steady for the past 10 years, despite interruptions, the link of the previous research interests to the new research interests could be presented in more of a story format so that the motivation for the past interest in memory development could be more clearly related to the current KO8 interest in reappraisals of trauma events following head injury in young adults with a history of treated relapsing drug use and alcoholism or no history of drug use and alcoholism. The primary mentor for the KO8 will be Dr. Julie Bethesda, who is well funded for research in preventing relapse in older drug users and alcoholics.

Strengths:

The candidate has a good history of training and publication and a willingness to pursue a career in science.

Weaknesses:

There is little in the self-description to tie the proposed research plan interests to the past history of research interests or to indicate how the proposed research and training will result in an independent scientist who is going to be more successful in obtaining grant support.

Career Development Plan

The 3-year career development plan shows coursework focused on pathways to drug abuse and assessment of mild cognitive impairment and dementia, which takes advantage of the Drug Abuse Institute at TTI as well as the Memory Center in the affiliated School of Medicine and area hospitals connected via telemedicine. There is also a focus on statistical analyses related to psychometrics and clinical trial outcomes. The mentorship plan involves attending grand rounds, a vascular dementia lecture series, and participation in head injury intakes in the emergency room, as well a working with the three area

(continues)

EXHIBIT 9.2. Sample Review for a Career Award (*Continued*)

associated hospital neurology clinics to screen and shadow head injury and stroke patients through their first 6 months of treatment and recovery. There is a timeline provided for expected progress, but there are no plans to monitor skills acquired at the various times provided in the timeline.

Strengths:

The career development plan takes advantage of the strength of curriculum and training at TTI.

Weaknesses:

The career development plan seems only loosely coordinated with the desired goal for independence and with the research plan. No timelines are offered for when skills might be acquired, or articles submitted.

Research Plan

The research plan is focused on assessing postinjury appraisal of head injury and motivation for recovery in two groups of young adults: those with and without a preinjury history of drug and alcohol abuse. The groups will be assessed for cognitive function; recovery motivation; and adherence to treatment for alcoholism and drug use; plus depression and anxiety at baseline, 4 months, 8 months, and 1 year after the injury. In addition to specific tests for executive function and speed of processing, the NIH Toolbox battery will be used. There is no history of the use of the battery by the mentors or the candidate. The statistical analyses focus on longitudinal change in function, plus survival analysis for any relapse to drug and alcohol use. No particular statistical analytic procedure is offered to cover the potential periods of differing improvement due to repeated assessment, dropout, or dropout and reentry into the study. There are general discussions of hierarchal linear analysis and multivariate approaches to analysis, but they are not tied specifically to the supposed interest in describing two different conditions, head injury and stroke. Preinjury use of drugs and alcohol is discussed as part of covariate analyses, but it is not clear how the covariate analyses will fit with the overall approach. There are several letters from the associate hospital neurology clinic directors verifying access to the clinic for recruitment, but the numbers that would be potentially recruited differ significantly from the largest to the smallest clinic, and it is not clear how many could be recruited from each site. It is also not clear how the history of drug abuse and alcoholism will be obtained other than from the participants and their family members. The letters of support indicate there are other trials being conducted at the clinical sites as well. Although training in clinical trials will be part of the career development plan, the proposed research does not show a clear arm of intervention other than those with and without a self-described history of drug and alcohol use. No pilot data are provided to support the methods proposed. The ambitious research plan is significantly underdeveloped.

Strengths:

No strengths.

EXHIBIT 9.2. Sample Review for a Career Award (*Continued*)

Weaknesses:

The research plan is loosely conceived and does not provide pilot data, evidence of literature support, or pilot data support for the premise. The methods are clearly undeveloped.

Mentor(s), Co-Mentor(s), Consultant(s), Collaborator(s)

The primary mentor, Dr. Julie Bethesda, MD/PhD, has a distinguished career in both psychiatry and neurology and has won a number of awards for her work in head injury and recovery from vascular stroke, including several grants, and is currently well funded. Dr. Bethesda provides a strong history of mentoring students, but only three have remained in academia or research. Several of her former students work in a research company started by her first postdoctoral mentee. The secondary mentor, Dr. Marsh (PhD, psychology), has been at TTI for 6 years and supported the candidate as a staff lab scientist for her past 3 years at TTI and was a student of Dr. Julie Ross at Old Allegheny. Dr. Bell and Dr. Brother, hired 2 years ago to direct the Vascular Institute clinical trials, are listed as mentors for statistical analyses and clinical trial development. The overall mentoring plan is vague, although the two primary mentors have adequate funding and a solid history of publication in the areas covered in the research and training proposal.

Strengths:

The mentors are well versed in their own fields.

Weaknesses:

Only one mentor (secondary mentor) has a significant knowledge of the candidate. The mentors write in support of the candidate but do not address how they will guide the mentoring process in any detail or assess progress toward independence or completing the aims of the research.

Environment and Institutional Commitment to the Candidate

The environments for the proposed research appear appropriate and are detailed with respect to research support, but it is not clear if the recruitment environments are appropriate, or where the primary commitment for lab space and support will be if the candidate receives the career award. The candidate is currently a career scientist supported in the lab of Dr. Marsh, the secondary mentor. Although the support letters from the supporting departments and the university are strong with regard to the strength of the candidate, it is not clear if the candidate must remain independently funded to achieve faculty status or if the award will provide that status.

Strengths:

The environment is appropriate for the research and has supported other successful candidates.

(continues)

EXHIBIT 9.2. Sample Review for a Career Award (*Continued*)

Weaknesses:

The status of the candidate as a research scientist applicant is apparently approved by the candidate's potential research department and the research institution, but the status of the candidate during or following the period of an award is not made clear.

Additional Review Considerations

(Reviewers will provide an overall impact score to reflect their assessment of the likelihood that the fellowship will enhance the candidate's potential for, and commitment to, a productive independent scientific research career in a health-related field, in consideration of the scored and additional review criteria. An application does not need to be strong in all categories to be judged likely to have a major impact.)

OVERALL IMPACT/MERIT:

The candidate provides an interesting history of development and is well regarded for her achievements in research and training. This strength is offset by the not-very-clear final goal of the K award and the lack of clarity in the training and research project focus. The mentors are favorable toward the candidate and the potential of a training award but otherwise seem disconnected from the project development process. The status of the candidate during the award or postaward time frame is never established.

Impact Score: 6

KEY TAKEAWAYS FROM CHAPTER 9

- For many research careers, an ideal training trajectory is to move from grant-supported predoctoral status to grant-supported postdoctoral status (if needed), to early (or late) career-supported status, or to jump from grant-supported predoctoral status to early-career–supported status.

- The goal of support is to have access to mentoring, financial support, research resources, and space for research and study. At the graduate school level, master's theses and doctoral dissertations are mentored experiences with access to available resources. Most funding agencies have mechanisms for dissertation that will support a specific set of aims and outcomes. The data from master's theses and doctoral dissertations can serve as pilot data for postdoctoral or RPG applications.

- Undergraduates often rely on mentors, who are supported through institutional training awards (T awards), to gain access to resources and mentoring, although undergraduates can obtain indirect support by enrolling in a research course or working in a graduate-level laboratory that is supported by federal grant money. Many granting agencies provide summer research experiences for undergraduates.

- Graduate students often obtain pertinent support by working as research assistants in a professor's laboratory but receiving course credit and anticipating paper presentation or publication credits instead of financial remuneration. Graduate students may be financially supported by a research grant support to a mentor by a training grant awarded to the institution and a group of researchers.

- Postdoctoral students working in the laboratory of a mentor have an elevated training and experience status. In some cases, postdoctoral students may work in a specific laboratory but are supported by research funds from an institutional training grant (T award) that was submitted by and funded to a group of researchers who are focused on a particular area of research.

- K awards (career awards) require evidence of prior research development. These awards require the most time commitment from candidates and mentors, and the training component is expected to develop a researcher who is independent and working on cutting-edge issues. Given the time period of commitment, changes in the stability of support from System 2 supporting institutions (department leadership changes, dean changes, mentors) are extremely important.

- Training awards are awards of commitment. The more advanced the training award, the more commitment is required from the applicant and mentor.

- Fellowship applications score well when the history of education and research can be developed as a storyline that supports the proposed research project.

- Career awards score well when prior experience melds well with future goals and there is a clear integration of training goals (often mentored) with proposed research aims. Career awards are not R01 applications with some training options added to the application.

- The NIH CSR has proposed changes to the review of fellowship applications that, if implemented would potentially focus the review's emphasis on the applicant's personal attributes and motivations; accomplishments will be reviewed within the applicant's existing and former training contexts.

10 SMALL BUSINESS GRANTS

Commercialization of Research

This chapter introduces small business grants to readers who are new to the business of commercializing research and provides further insight for those who have already been competing for such grants. Often, advanced knowledge in an area can be further consolidated by revisiting the basics once one has gained an enhanced perspective. Small Business Innovation Research/ Small Business Technology Transfer (SBIR/STTR) mechanisms fill a major gap in funding for inventions and new products. These are federally sponsored mechanisms and subject to federal funding and federal policies. The Bayh–Dole Act (Pub. L. 96-517, also known as the Patent and Trademark Law Amendments Act), which is discussed in more detail shortly, is one of the federal policies that directs the ownership of federally funded products. The Bayh–Dole Act describes the reporting requirements for inventions developed with federal funds; knowledge of this legislation is essential if one wishes to retain ownership and control of the product.

If applicants/entrepreneurs are introduced to federal research applications and to small business applications at the same time, they may need time to acquire the knowledge needed to integrate research and commercialization

https://doi.org/10.1037/0000390-011
Get Funded: A Practical Guide to Understanding the Grant Application Process and Writing Winning Proposals in the Behavioral and Biomedical Fields, by J. W. Elias

skills. Applicants who are knowledgeable about research and product development and prepared to develop an SBIR/STTR proposal should know that it can take 6 to 8 weeks to respond to a funding opportunity. They should be aware that SBIR/STTR funding often develops in phases that initially fund proof-of-concept research (see https://www.sbir.gov/about). From the start, researchers/entrepreneurs should know that limits on Phase II funding may not provide all the resources needed to solve methodology issues discovered in the review process. When federal grant funding became a more dominant form of advancing science, the Bayh–Dole Act permitted universities to patent the products of science that had been achieved through contracts, grants, and cooperative agreements. The legislation emphasizes the importance of managing intellectual property developed from federally funded research but does not require investment in the development of the property. Universities could own the titles to patents that would then permit licensing of products ("What Did the Bayh–Dole Act Do?," 2022). The Bayh–Dole Act permitted the federal government to promote important products of grant funding while allowing universities and nonprofit organizations to retain ownership.

To accelerate science and marketable products in 1983, federal agencies with an extramural research and development budget were directed by the Small Business Innovation Development Act of 1982 (Pub. L. 97-219) to spend a designated percentage of their budget on an SBIR program This congressionally encouraged program was designed to help scientists develop their products to the point at which there was a potential for commercialization. Reviews of the program indicated that it was successful (see https://incumetrics.com/national-academies-small-business-and-innovation-studies/), and congressional authorization for the legislation has been extended and reauthorized several times since its initial enactment in 1982.

In 1992, Congress approved the STTR program. This action encouraged and enhanced joint research and product development collaboration between small businesses and nonprofit research institutions. To use the STTR mechanism, a small business must collaborate in a percentage-of-effort fashion with a research institution. Under current rules, the nonprofit research institution partner must contribute 30% to 60% of the effort.

As stated on the SBIR/STTR America's Seed Fund website (https://www.sbir.gov/about), the STTR differs from the SBIR in three ways:

1. The small business awardee and its partnering institution must establish an intellectual property agreement detailing the allocation of intellectual property rights and rights to conduct follow-up research, development, or commercialization activities.

2. The small business must perform at least 40% of the research and development, and a single partnering research institution must perform at least 30% of the research and development.

3. The STTR program permits a Principal Investigator (PI) to be primarily employed by the partnering research institution.

In 2016, the SBIR and STTR programs were approved through September 30, 2022, by the National Defense Authorization Act for Fiscal Year 2017 (Pub. L. 114-328). In 2022, there was a last minute Congressional approval and signing by the President of the SBIR and STTR Extension Act of 2022 (Pub. L. 117-183; see Gill, 2022). The new 3-year authorization was delayed by discussions regarding foreign influence on SBIR/STTR funding and the overuse of companies using the Phase II funding program without ever progressing to Phase III. The phase components of SBIR/STTR proposals are discussed in detail later in this chapter and can be found online (https://www.sbir.gov/). The Small Business Administration, as a federal agency, coordinates the SBIR/STTR programs for federal agencies that distribute the funds as part of the granting process.

New and experienced applicants/investigators/entrepreneurs interested in accessing SBIR/STTR funds should not be surprised by the continual changes in the SBIR/STTR programs, which are due to the need to reauthorize the funds via a political process. In addition to general changes in SBIR/STTR programs, each funding source has its own perspective on funding applications and offers Funding Opportunity Announcements/Notices of Funding Opportunities (FOAs/NOFO)s that indicate how the SBIR/STTR program fits with the interests of the funding source. The top five federal funding sources for the SBIR/STTR program are the Department of Defense (DoD/DOD), the National Institutes of Health (NIH), the National Aeronautics and Space Administration (NASA), the National Science Foundation (NSF), and the Department of Energy (DOE). These science-commercialization programs typically involve national competitions but are proof-of-concept driven by economics and by international competition for advancements in defense, health, agriculture, computing, and basic science.

Despite the fact that small business programs have existed for close to 40 years, many of the details are distributed across multiple federal programs to support entrepreneurship and the commercialization of scientific findings. The multiple agencies that support small business programs with grants and contracts have improved their outreach. Several gateway .gov websites have been updated that are worth exploring for their directions on how to develop an SBIR/STTR proposal. An internet search for "SBIR funding opportunities" will lead to several websites that offer relevant information. Any source of information on the internet should be verified for its timeliness. Sites with a .gov extension provide reliable and free information, but one should always check the dates of the information.

The term "Seed Fund" is used by the Small Business Administration (https://beta.www.sbir.gov/about), NSF (https://seedfund.nsf.gov/), and NIH (https://seed.nih.gov/) to discuss the entrepreneur-type programs.

The basic information needed to understand small business programs is available through the .gov sites, which also provide information on specific funding agency interests, science expectations, and business expectations. Practical housekeeping information can be found at these sites, including applicant eligibility requirements; the benefits of the SBIR or STTR approach; how SBIRs/STTRs can develop in phases (i.e., Phase I, Phase II, FastTrack, Direct to Phase II, Phase III); how much time the first phase, or multiple phases of funding, might require; the financial support allowed at each funding phase; and ways to find knowledge-based support when starting out as a novice.

Unlike regular research applications, which can request funding for support to complete a project, or for additional support to add on to an original grant, the SBIR/STTR Phase III commercialization phase is not funded, but information from funding sources is available on how Phase III can be completed with help from the business community. SBIR/STTR businesses may not be permitted to submit as many Phase II applications as needed to address reviewer concerns that keep scores above the funding lines. Businesses that continually submit and are funded for Phase I projects that never move to Phase II projects may be limited in further submissions. The goals of SBIR/STTR federal authorization are that submitting and receiving funding should result in Phase III commercialization. The business should not simply receive funding for Phase I and Phase II.

SAMPLE APPLICATIONS PROVIDE GOOD GUIDELINES FOR DEVELOPMENT/COMMERCIALIZATION

The National Institute of Allergies and Infectious Diseases (NIAID) provides examples of older applications (https://www.niaid.nih.gov/grants-contracts/sample-applications), but the principles of the application remain the same. The examples allow one to compare the focus and detail of Phase I (proof of concept), Phase II (strengthening the proof of concept and a commercialization plan), and a FastTrack application that combines Phase I and Phase II.

The sample applications provide a good example of how a small business seeking funding should be thinking about their product and its ability to be developed and marketed via an SBIR/STTR application. The amount of information found in the examples can be a bit overwhelming for newly initiated SBIR/STTR applicants, but the submission processes are systematic and can be managed with persistence and patience. Access to the sample applications permits quick insight into how SBIR/STTR applications differ from R01 applications. Phase I applications look quite a bit like R01 applications in terms of the sections of the application and the review process. Phase II applications include the information gained from the previous Phase I application, explain further development of the scientific components of the product, and discuss a commercialization plan. Potential SBIR/STTR applicants should examine the components of the commercialization plans of sample funded applications.

Commercialization plans will include the value of the developing product relative to similar products, the expected outcomes of the development research, and the impact of the outcomes in the marketplace. The commercialization plan will outline how the company will produce and market the product and describe the plan for marketing. Finance plans and sources of revenue over time will be important. Commercialization plans also provide descriptions of the targeted market for the product, the range of customers, and descriptions of competitive products already on the market or in development. The application forms continually evolve, so the most recent formats should be used to guide application submissions.

COMPARING THE SBIR/STTR WITH THE RESEARCH PROJECT GRANT

Scientists moving from a Research Project Grant (RPG) grant proposal to an SBIR/STTR proposal will notice several differences immediately:

1. The goal of an SBIR/STTR grant is to support a partnership between research and business to produce a marketable product whose worth has been proven technically by a scientific method.

2. The partnership mix is between a business component and a research component, and the percentage of employment by the Program Director (PD)/PI determines whether an application is an SBIR or STTR. The PD/PI in an SBIR must be employed at least half-time by the small business, and the relationship (unless waived) should remain that way through the phases of product development.

As previously discussed, an STTR proposal requires a formal relationship between the small business and a nonprofit research institution. This relationship requires that the PD/PI and the small business must contribute at least 40% of the work through the phases, and the research institution must contribute at least 30% of the work effort. The remaining 30% can be divided between the small business and the research institution or a third party. The percentages of work are reflected in the direct and indirect cost percentages stated in the budget. Universities can work in collaboration with the business entity but cannot apply as an applicant.

3. Eligibility for an SBIR/STTR grant is a key component of the decision to develop an application. Applicants should read the policy rules regarding ownership and eligibility and should not assume, even with prior experience, that they are up to date with any new policy. Applicants' percentage of ownership of the company is important, and principal owners must have at least permanent U.S. citizenship or permanent residence status. The percentage of the business owned by subsidiaries can be important because the goal is to fund small businesses (no more than 500 employees). Eligibility extends beyond citizenship status and company ownership and size to include the ratio of Phase I awards to Phase II awards. This ratio is designed to make sure the company business is not submitting Phase I applications that never succeed beyond that phase. The recent authorization of the SBIR/STTR program notes that there are concerns about companies that receive multiple Phase II applications but do not take the product the commercialization phase. The goal is commercialization.

4. Applications are submitted in response to a soliciting SBIR/STTR announcement. Not all federal agencies fund SBIR/STTR grants; a general search for funding agencies will lead to specific agencies, such as NIH, NSF, DOE, and DoD/DOD. All the funding opportunities will require specific registrations with those agencies before submission. Research institutions have knowledge of these registrations, but this information may be new to small businesses.

5. SBIR/STTR grants typically develop in phases, such as Phase I, Phase II, and Phase III, and the federal funding dollars and time allotted for each phase increase from Phase I to Phase II. Phases can be put together and fast-tracked, or a Phase II application that encompasses a Phase I application can be proposed as a "Direct to Phase II" award, which is used when the information needed from a Phase I application has already

been acquired. Phase III (commercialization) applications are not funded by federal funding agencies. To offset the leap to Phase III, and the move to provide financial support to investors, some agencies' Program Announcements allow an extended Phase II application to be submitted to provide further funds toward establishing or extending a commercialization plan. One must sort through the SBIR/STTR FOAs/NOFOs to find these opportunities.

The goal of the granting process is to progress through the phases to the point where commercialization seems possible to investors. *Commercialization* means that there must be a market and a marketing plan that indicates what the competition might be and the advantages of a new product. The funding gap between Phase II funding and Phase III commercialization is often the most difficult part of the production process. At this point, companies should scour the FOA/NOFO, Research Funding Announcement (RFA), Program Announcement with Special Review Criteria and/or Specific Receipt Dates (PAR), and Notice of Special Interest (NOSI) for information about accessing help with commercialization.

6. Federal agencies are aware of commercialization difficulties and have developed programs to help funded grantees understand the process and to connect with investors in what is referred to as *Phase III* of SBIR/STTR development (see https://seed.nih.gov/small-business-funding/find-funding?redirect_from=sbir.nih.gov#crp). There is a nascent interest in training early-career faculty and postdoctoral students to be entrepreneurial scientists by means of the SBIR/STTR mechanisms. The National Institute on Aging (NIA), for example, provides entrepreneurial training opportunities and promotes its interests in training for entrepreneurship (https://www.sbir.gov/node/2234825).

> A key difference exists between this NIA program and other SBIR awards in that this program is specifically aimed at early career scientists who are interested in entrepreneurial training and mentorship. Additionally, this award provides an opportunity for early career scientists to grow their leadership skills while serving as a PD/PI/RFA. (https://grants.nih.gov/grants/guide/RFA-files/RFA-AG-23-029.html)

There is a growing recognition that pre–Phase I activities might be needed to encourage investors to take an early interest in emerging business–research ventures from small businesses; that is, part of Phase I is to start planning for the commercialization that takes place in the non-agency-funded component of Phase III. As noted, NIH has developed a SEED office. Programs such as the BIO Innovation Zone (https://www.nsf.gov/news/news_images.jsp?cntn_id=138773&org= NSF) have been connected to the program. A recent development is a Company Showcase program, operated in partnership with the SEED Office and NIH institutes' and centers' SBIR/STTR program leadership. Applicants should be encouraged to attend presentations and seminars on entrepreneurship to learn about protecting potential products via the patenting process and to be aware of how public presentations of potential product and methods can affect the patenting process.

7. For applicants/investigators/companies that are new to the grant-writing experience, the skills required to apply and receive funding include technical skills, statistical skills, grant-writing skills, business skills/business acumen, fund-raising skills, and post–grant-management skills that go beyond product development. These skills do not need to be mastered by just one person, but the skills within the business should cover the functions the business. For companies that have not had NIH funding experience, it is prudent to include in their planning how they will manage the postfunding grant paperwork and policies according to the funding agency's guidelines.

8. Product development reality is a major component of the small business venture. For example, if a company developed a new kind of pencil and eraser, and they were interested in maintaining this product and using all of its components, the reality might be that only the eraser is drawing the interest of investors.

9. There are risks involved that may not be involved with research grants, such as

    – short- and long-term liability issues with respect to the product;

    – the costs and risks of maintaining a business that are not required for maintaining a laboratory;

    – risk management for employees;

    – a product that may have to be further developed to maintain a market edge or viability;

    – products with a short shelf-life in terms of viability, requiring developers to think about both the immediate competition and then what might be developing;

    – the reality that managing a supply chain can be a full-time job; and

    – understanding that statistical significance is important but that it does not have the same impact as stating it in an article; finding overlapping statistical distributions can have multiple meanings in product development and businesses are not required to continually perform statistical analyses once a product is developed.

10. Although often overlooked, the choice of a funding agency is the initial investment decision. Funding agencies view grants that are funded as investments, and different funding agencies have different funding interests, funding policies, and budgets for SBIR/STTR grants. The application focus and the review processes guide and limit the choices funding agencies can make, but investors would do well to examine the funding agency's budgets, plus the number of projects awarded and the assistance offered relative to commercialization. Obviously, the nature of the product can limit the range of investment partners at funding agencies, but it is still wise to carefully consider the first funding decision after talking to agency Program Officers (POs). Keep in mind that some POs have more experience with and interest in SBIR/STTR grants than others. Some institutes have significantly increased the percentage of grant dollars allotted to SBIRs/STTRs.

HOW TO USE THE NIH DATA BOOK TO FOLLOW FUNDING TRENDS

The NIH RePORTER and Data Book provides yearly reports on NIH institute funding by mechanism and indicates the number of projects received and the number funded. Table 10.1 shows the number of applications for SBIR/STTR grants submitted and funded by phase of mechanism across the years 2018–2022; the data are compiled from NIH funding reports that can be found at https://report.nih.gov/nihdatabook/ca. It is assumed that STTRs are represented in the SBIR data given that the top banner of this report includes SBIRs. The drop-down menu on the sidebar of the site shows where SBIR/STTR data can be located.

In addition to Direct to Phase II applications, most Phase II applications must pass through the Phase I funding process. The Phase II NIH Data Book posted success rates are calculated on the basis of the number of Phase II applications submitted versus those that are eventually, but not necessarily, immediately funded. In other words, the success rate of Phase II applications does not precisely represent the pass-through process that requires successful Phase I funding.

Table 10.1 is designed to depict this step-down process and shows, as expected, that from 2018 to 2022 there were more Phase I submissions than

TABLE 10.1. SBIR and STTR Grant Applications Submitted and Funded by Phase Mechanism

SBIR/STTR application phase	2018	2019	2020	2021	2022	M
Applications submitted (*n*)						
Phase I	3,707	3,513	3,630	3,971	3,705	3,705
Phase II	618	774	1,260	1,430	1,021	793
FastTrack	731	665	637	682	679	679
Applications funded (*n*)						
Phase I	665	671	450	450	561	561
Phase II	257	258	324	324	286	285
FastTrack	148	159	108	108	129	129
Applications funded (%)						
Phase I	18	19	12	12	15	15
Phase II	42	33	26	21	26	30
FastTrack	20	24	17	15	18	19

Note. SBIR = Small Business Innovation Research; STTR = Small Business Technology Transfer.

Phase II or FastTrack submissions. Phase II applications funded in the years 2020 and 2022 were likely from a pool of resubmitted Phase II applications from the previous years (2018–2019) or a pool of funded Phase I applications funded in 2018 and 2019. The actual percent funded Phase II comparisons is a function of the percent funded Phase I applications. Eighteen months is the likely minimum time required for applicants/investigators to develop a Phase I application, submit it, and then be Phase I reviewed and funded and completed. Applicants/investigators would then require a minimum of 3 months to submit the following Phase II application and another 9 months to receive news of funding for a Phase II application.

The success rates of Phase II applications do not represent the actual compound probability of being funded from the point of Phase I given that most Phase I applications are not resubmitted in continuance as Phase II applications. The probability of a successful Phase I application, multiplied by the probability of a Phase II application being funded, is a very rough estimate of the compound probability of final Phase II funding because the data points used for Phase II funding would represent only some of the same applicants from a specific Phase I funding year. From my perspective, the ability of SBIR/STTR applicants to submit a FastTrack application with strong initial proof of concept data that allows a combination of Phase I and Phase II greatly reduces the months before final funding (or resubmission) and is a more viable model in terms of time of development comparable to a direct to Phase II application.

NIH reports that for 2021–2022, the overall success rate for grants categorized as RPGs was between 20% and 22%. which puts the challenge of receiving SBIR/STTR funding into perspective (Lauer, 2022a, 2022b). SBIR/STTR funding rates do not take into consideration how many of the products are successfully developed and marketed, just like the funding rate of RPGs cannot predict how many of the funded applications can be considered a success. Nevertheless, estimating the *time of development and success rate* for Phase II applications requires that one consider the preparation and pass-through process from the initial Phase I application. According to the aggregate statistics provided by the NIH Data Book and the data presented in this text in aggregate form in Table 10.1, a successful Phase I submission predicts a high success rate for a subsequent Phase II submission.

SIMILARITIES AND DIFFERENCES/CONCERNS IN THE REVIEW OF RPG AND SBIR/STTR APPLICATIONS

For applicants, the structure of the overall review process is similar for RPG and SBIR/STTR grants, but the SBIR/STTR focus for reviewers and review panels is extended beyond science methodology to include business methodology and entrepreneurship. Given that more domains of comment exist for small business development, the scores that the comments represent are likely to be more variable across both science and business dimensions because the review panel members are recruited for their representation of science, business, and entrepreneurship.

SBIR/STTR reviewers are often conservative in perspective even though the applications focus on new and potentially innovative concepts. It is not unusual for reviewers to show resistance to further product development if there is a perceived need for more data to show viable product usability. Projects that involve risk to the users of the new technology are more likely to receive cautious reviews. This is a difficult situation to navigate if there are no more funds available through Phase II funding or other competitive funding sources to continually collect more data to satisfy reviewers.

When applicants/entrepreneurs may have to find non-SBIR/STTR funds to move the project forward, this means further negotiation about the owners of the data and patent and licensing issues. When projects involve some risk of adverse events to consumers, the best approach is to assume reviewers might take a conservative stance on methodological issues and plan to show proof of function beyond that required for funding typical RPGs, such as R01s.

A Phase I application designed to have smaller funding amounts and limited time should be carefully crafted to overcome the most likely technical issues to be encountered when the Phase II application is submitted. A best practice would be for applicants/investigators to sketch out what the major issues in Phase II would be. It is unfortunate if applicants/investigators come up short in resolving technical issues in Phase II because there is a limited budget and a limited Phase II time frame in which to overcome sticky technical issues. Applications with a higher risk to subjects may require additional demonstrations of safety and applicability.

For applicants/investigators, the expectations for System 3 funding support can extend from Basic Research (aka Phase I) to Invention (aka Phase II), and possibly some help with Technology Promotion/Marketing (aka Phase III), but investors outside of System 3, such as private investors, corporations, and venture capital investors, are expected to fund technology to the point of product development that then leads to product marketing. Applicants/investigators who move into Phase II funding hoping to entice Phase III funding start moving toward what the entrepreneur-based industries have called the "valley of death" for technological development if Phase III support cannot be found. Successful Phase II applicants can receive assistance from federally contracted programs designed to move projects toward Phase III.

For SBIR/STTR applicants who are familiar with the typical process of developing the Aims, Significance, Innovation, and Methods sections, there is a definite advantage to knowing not only the organizational components of a proposal but also how to navigate the systems involved at the initial level of federal funding. However, these sections have more work to do than in an RPG. These components must support proof of concept/invention, product development, and a vision for commercialization and a viable business plan.

Unlike research grants, where the reviewers are often told to stick to evaluating the science and the impact of the science on advancing the field (and funding agency POs decide on the funding), even in Phase I the SBIR/STTR review panel is focused on the potential for the project to produce a stream of revenue. Part of establishing, as an investigator/developer, how good an idea is means searching for how many others think it is viable by searching patents and looking for funded grants that have promoted similar ideas.

SBIR/STTR applicants need to establish in the Aims section that a viable product is needed as part of the solution to a scientific problem that will be solved. The premise supports the scientific approach and the significance of the product. The Methods section provides strong proof of concept development. Applicants' products are expected to be innovative and add value to existing products and methods or provide value that is new. Even innovative products introduced into a crowded market may be received with some skepticism that the new product can establish a market share. Exhibit 10.1 lists several of these "pay-attention-to-my product" issues.

There is a caveat to introducing the advances of a product in an SBIR/ STTR application. Reviewers and potential users will want to see how the advances promoted will be defined and how the promoted advances will be proven. The promoted advantages should be carefully gauged for their believability and their ability to be reliably judged within the scope of the application, including its budget. There may be six ways in which a product will induce or introduce positive change or will be better than an existing product, but there may be only three ways these claims can be proven. From a science perspective, proof often resides in statistically significant outcomes as a start. Applicants need to address issues of reliability and replicability and move beyond statistical significance to define effectiveness by the percentage variance accounted for or the degree of overlap in comparisons of groups. For example, small but reliable changes in the usability of a product touted as a selling point might not be worth a difficult change in habits being asked of the product's users.

When promoting the advances of the product, applicants should consider the feasibility of the approach. The greater the need to establish or explore issues of credibility or feasibility and proof of concept, the greater the need for pilot data. Phase I is often the pilot data production mechanism. The presence of strong feasibility data helps applicants decide whether to choose

---

EXHIBIT 10.1. Selling Points in an Application for Small-Business Funding

- High compliance with use—simple, easy, and "good enough" can be major selling points (e.g., quick development and sharing of pictures)
- Low or reduced error rate
- Limited maintenance or reduced maintenance, not highly regulated
- Efficient local access
- Increased user/patient confidence; there is a preference for the product
- Reduced risk of injury
- Privacy is secured
- Little need to change habits with the new product, or new habits for use can be easily formed
- The product meets little resistance to change in several dimensions and scores well in readiness to change by potential users
- The product allows multiple processes to occur within the same time frame as other processes and thus saves time and effort

Phase I or Phase II, or a FastTrack Phase I–Phase II combined. When Phase I shows acceptable feasibility, this should be made known in the resubmission of a Phase I application or in a subsequent Phase II application. Applicants should keep in mind that the funding process is a competition and providing enough data to establish feasibility may not be competitive with other applications who have data to show strong feasibility.

As with a fellowship award or a career development award, applicants new to the SBIR/STTR granting process will need to find experienced mentors to help write and produce applications. Applicants coming from the research side, where experience in writing grants is high, should focus on acquiring into System 1 resources, including colleagues who can provide technical assistance for product development and, ideally, contacts with business partners. The System 1 (or System 1–System 2) stress points for a business-naïve group will be in developing the production and business team as opposed to understanding the basic guidelines for submitting a grant proposal. Having experience in writing grants or access to grant writing resources as part of the System1 and 2 components will allow applicants to make a more accurate estimate of the time required to develop a written document, submit it, and wait for initial reviews.

Functional Integration of Business Management and Science

A common review comment for SBIR and STTR funding is that the proposed activities are ambitious. Frequently, that comment indicates that the reviewer thinks the applicants are not capable or that the project as described is not feasible. Applications for SBIR/STTR grants are ambitious projects and need to be innovative because they are going to promote change. Therefore, reach should be right at the point of grasp.

A component that applicants overlook in describing SBIR/STTR management is how the management will interact and solve problems. One way to ensure that there is capable management of the complexity of the SBIR/STTR application is to provide a detailed description not only of who is part of the company and their designated roles and experience but also of how the individuals in the company will interact to provide the proof of concept and the marketing plan for the expected product. An excellent source of information about how to describe the administration of an SBIR/STTR grant can be found by looking at the NIH Program Project grant (Activity Code P01). The P01 requires a discussion of how multiple projects can be managed and asks for organization information that would be helpful for SBIR/STTR applicants who wish to describe how their business will be organized and managed.

NIH defines a Program Project Grant as follows: A Program Project Grant (P01) is an assistance award for the support of a broadly based multi-disciplinary research program that has a well-defined central research focus or objective. It may also include support for common supporting resources (cores) required for the conduct of the component research projects.

In an SBIR/STTR application, the multiple goals and timelines must be managed. In the next paragraph are listed some of the details required for an administrative core exactly as presented in PAR-21-181 (https://grants.nih.gov/grants/guide/pa-files/PAR-21-181.html; scroll down to "PHS 398 Research Plan [Administrative Core]").

When using the objectives outlined for the Administrative Core in a Program Project (P01) application, the following organizational components should be addressed in an SBIR/STTR:

- Introduce SBIR/STTR staff (including administrative staff) and their role; if staff are added sequentially, note the time of their addition.

- Explain how SBIR/STTR administrative/staff roles and function fit with the goals of the SBIR.

- Describe how SBIR/STTR staff and administration interact and how often they interact with respect to planning and coordination.

- Indicate how strategic issues are determined/managed by SBIR/STTR administration and how disagreements are managed.

- Identify how funds are allocated by SBIR/STTR administration to maintain the integrity of the proposed project.

- Describe the internal review processes that will be used to assess SBIR/STTR management communication and quality control.

- Indicate whether SBIR/STTR advisory committees will be established and how and when they will interact with administration.

- Describe the visual organization of the SBIR/STTR administrative roles and the nature of the interaction.

Market Resistance in Reviewers

Market resistance to a product can occur in unexpected ways, and this is important to address early in promoting the SBIR/STTR application in Phase I and to do so completely in applications involving Phase II. Reviewer panels are a part of the product development process, and they will be the first in the line of product development to reveal resistance to the product, either for themselves or by thinking about it in relation to the potential customer. The more data that can be provided to reviewers about relative market readiness for a product, and the ease with which customers are likely to adopt new ways and habits, the greater the likelihood of eliminating points of potential product resistance. Reviewers are typically on the cusp of thinking, "This product will probably be of interest to investors and the public" or "This product will probably not be of interest to investors and the public."

Because SBIR/STTR grant applications are peer reviewed, applicants should endeavor to convince the reviewers (not just the potential investors) that the product is important. Given that most grant applications are not funded on the first submission, the Summary Statement should be used to gauge the degree of market resistance shown by reviewers even if the resistance is cloaked in a series of small complaints and concerns.

The possibility that there is a risk to using a product can often be the tipping point for a reviewer in terms of personal concerns and resistance, and any potential risk should be addressed as thoroughly as possible. The issue of gains and losses and the willingness to take a risk to avoid a loss, or to take a risk to make a gain, has been addressed for many years in the literature on economics and the marketplace (e.g., Kahneman & Miller, 1986). The literature is replete with how decisions are made relative to risk. Reviewers will be making these decisions initially, and if they do not give the product a positive review then there will be no SBIR/STTR dollars to support getting the product to the investors or consumers. Therefore, a critical area of research for SBIR/STTR funding applicants is the decision process that is used to accept risk in an area of product development. For example, applications applied to biological development processes will be perceived through a reviewer and investor lens of long-term consequences and gain. Procedures designed for young people and older adults will have different long-term consequences.

PERSPECTIVES ON SBIR/STTR FUNDING

The complexity of SBIR/STTR funding stems from the number of critical parts of the mechanism, including the need to show evidence of strong science, strong management, strong business skills, an emerging market, and a plan for reaching that market. The fact that SBIR/STTR funding can be developed and competed in stages is a clear advantage to reaching the point at which investors step in.

When companies propose innovative products, the emphasis is on presenting the product or products in terms of their maximum effectiveness for utility and for acceptance in the marketplace. Entrepreneurship involves promotion and enthusiasm for the final product and its niche in the marketplace. Public presentations of ideas and potential outcomes can provide positive feedback and motivation. Moving to SBIR/STTR funding to finance the early stages of product development is a good choice if eligibility requirements for small business competitions can be met. It is best to assume from the beginning of the grant application process that the reviewers will be providing reasons or asking questions regarding those two primary selling points, that is, utility and acceptance in the marketplace.

KEY TAKEAWAYS FROM CHAPTER 10

- The SBIR/STTR program is one of the most competitive funding programs because small business grants share, with research grants, a need to support the underlying science and must develop product commercialization and commercialization plans. The patenting and potential licensing of products are important concepts to be addressed when beginning to develop a product that might have commercial value.

- The SBIR/STTR programs are programs of discovery, translation, application, and ownership. They have commercialization and marketing components that set the programs apart from applying for a patent based on a discovery related to basic research (e.g., an R01 produces patentable product). The best way to see the nature of SBIR/STTR programs is to access the samples provided of funded and review applications on funding agency websites.

- Three support programs with the acronym SEED have been offered by the (a) federal government, (b) NIH, and (c) NSF to help navigate the components of the SBIR/STTR application process (https://www.sbir.gov) and the processes of the SEED program within the funding institutes (e.g., NSF [https://seedfund.nsf.gov/] and NIH [https://seed.nih.gov/]). These websites are designed to help applicants/entrepreneurs navigate the SBIR/STTR application process.

- SBIR/STTR programs typically develop in phases. The Phase I component of an SBIR/STTR application provides smaller funds and a short period of time to develop proof of concept. The next phase (Phase II) expands the proof of concept to application and commercialization potential. The Fast Track option, if possible, provides the fastest time to funding without reducing the probability of funding.

- The reported success rate of a funded Phase II proposal that starts from a submitted Phase I proposal does not represent the overall success rate starting from Phase I. The true potential of Phase II funding must take into effect the initial funding potential for Phase I applications.

- In addition to developing plans that integrate science, product development, and business development, SBIR/STTR funding success depends on the successful management of these multiple components. Applications that detail the way the components of development are managed can help convince reviewers and potential investors that the production, application, and marketing of the product are feasible.

- There are common review issues that SBIR/STTR applicants should plan for. Some of these concerns are methods based, some are market based. Both kinds of concerns can be anticipated.

- When market acceptance requires accepting some level of risk to consumers, then stronger data will be needed to convince reviewers and funders that market need is high and usability is clearly established.

- An examination of FastTrack data indicates that a FastTrack application either represents greater readiness, accounting for the 17% approval rate, or that reviewers like to see the justification and the commercialization plan together in one package, which also represents greater readiness.

PART V ADVANCED GRANT WRITING

11

TIME AND TEAM COMMUNICATION MANAGEMENT STRATEGIES

Grant-dependent positions require steady access to data; they also necessitate not only financial support but also temporal support as resources to develop multiple sources of grant funding. Time is always at a premium in academic, academic–clinical, and academic–research environments.

System 1 applicants attempting to submit an application by a specific date quickly learn that one of the most important skills is time management and one of the most important resources is time to process. Time management is a concern for new, midlevel, and advanced investigators seeking grant support. Positions that are grant dependent require research skills, the ability to work in a research team atmosphere, and the capability to develop time management skills.

It is human nature to focus on the more interesting aspects of grant writing, such as developing ideas into aims and hypotheses. It is also human nature to underestimate the time required for the initial writing of an application and the number of revisions and editing that will be required if the application is to be competitive. Nevertheless, there are, for new investigators in particular, several components of grant submission that must be mastered in addition to writing. The number of components for a submission to large

https://doi.org/10.1037/0000390-012
Get Funded: A Practical Guide to Understanding the Grant Application Process and Writing Winning Proposals in the Behavioral and Biomedical Fields, by J. W. Elias

funding agencies seems to be multiplying, particularly in regard to policy, regulation, conflict of interest, human subjects, budget, Data Management Plans, letters of support, forms, rules for including content, and rules and timelines for submission on both the local and agency levels. Over the years, I have observed many grant applications scored poorly, returned without review, or not submitted because of poor time management that resulted in poor writing and editing or incomplete ideas.

The first step toward effective time management is to develop a timeline. A timeline should include the time required to understand and develop all the components of a grant submission. If an investigator does not have sufficient time, or is too impatient and disorganized to develop a timeline, then they are already leaving some components of competitive grant writing to chance. If applicants are not familiar with the components of a submission, the time-line should include an appropriate amount of time to become familiar with the needed components. Investigators who are familiar with the components of a submission recognize that leaving some components—such as letters of support, regulation compliance, and budget development—until the last few weeks before submission takes time away from an essential component of grant writing: revising and editing the final document. This is where even experienced investigators can make themselves less competitive without recognizing the problem.

Step 1 in time management is to find out what must be accomplished every time an investigator submits, because new rules and policies appear, and greater emphasis is placed on components of applications that used to be considered additional material.

Step 2 in time management is identifying the individuals at one's institution who constitute the support team for submitting applications and finding out, early, what is required to process an application and what the timelines are for processing.

Step 3 in time management is to recognize that the last date for completing a component of the process may not be open if unforeseen circumstances arise and if there are too many applications competing to be processed at the last moment within System 2. Just because an office is open until 5:00 does not mean you can have your business processed at 4:50, or even on the last day of a due date.

Step 4 in time management is to clarify who is responsible for Steps 1 through 3.

System 2 management policies play a role in time management. Through work policies, System 2 management determines the structure of how time can be spent in the workplace. In this sense, System 2 can distribute time for projects such as grant proposals. The broad functions of some System 2 institutions may not permit much opportunity to provide flexibility. The more flexible System 2 institutions may have an advantage in supporting grant applications. System 2 administrators should be aware that frequently they are the timekeepers and the time-givers in a research institution. They have a budget to distribute that may not directly fund time but budgets indirectly translate to the time one has available. Systems 1, 2, and 3 budgets are driven by agendas, and time is a major component needed to complete any agenda. A less well-recognized and developed administrative skill is the ability to plan for the time a budget may need to produce fruitful outcomes. In societies that tend to focus on more short-term goals and agendas, there may be little reward for playing the long game as an administrator if the goal is to move either oneself and/or an institution upward and onward. Some research programs and some researchers do not fare well under these conditions. Experienced System 1 investigators and System 2 administrators plan for streams of funding with the knowledge that there will be gaps in grant-funded support.

Investigators who transfer grant proposals from one institution to another should expect at least 1-year delay if data collection must move locations as well. Investigators who move grant resources and assume the role of administrators should plan for 2 years of disruption in data collection. It takes time to organize new laboratories and support elements. Recruiters always provide the rosiest of outcomes for establishing laboratories. Investigators will find there is little benefit in delaying the initiation of paperwork to transfer a funded grant to a new institution. Principal Investigators (PIs) cannot play a passive role in transferring an application. It is the PI's responsibility to make sure the process is moving on from the old institution and to keep track of agreements to process paperwork in both the old and new institutions, and to monitor timelines and communication. Program Officers (POs) are interested parties but are typically not monitoring the process other than by contact with the PIs.

TEAM DEVELOPMENT, MANAGEMENT, AND EVALUATION

For applicants/investigators, the grant application is an attempt to communicate ideas across a technological grid for the purposes of making a social connection with a review group, who will reciprocally provide a social perspective on the value of the proposed research in the form of a Summary Statement. Very few grant proposals are accepted without feedback that requests or suggests some change in the proposed research. Therefore, the science proposed by applicants/investigators is altered by the ability to communicate clearly and socially on the part of the applicants/investigators and the reviewers. For grant applicants who are not socially adept at communicating or comprehending ideas via written or verbal communication, having a good team that can translate is essential. Communication skills can be viewed as a resource for both individuals and research teams.

For each System 2 institution, the functioning of the team and the number of team members, and the complexity of the team process, may differ. The better the research team's understanding of the full process and, very importantly, the time frame of each component of the team process, the better the team members will be able to work together. The bottom line for team members is that they should endeavor to understand more than just their specific research component. Team leaders should provide team members with an outline of a process the PIs would like to follow.

For many grant proposals for which there is only a brief period to respond to a specific funding opportunity and assessing the functioning of team performance is done on the fly, during the development of the application, knowing where problems may occur is useful. Experienced team leaders and team members know that team projects will develop at a different pace depending on the specific team members and the task at hand. A classic article discussing the potential stages of project development was written by Tuckman (1965), who described stages of (a) getting acquainted; (b) discovering and communicating individual perspectives; (c) joining together to meld individual perspectives; (d) working together toward a goal; and (e) as the project comes to an end or is completed, team assessment of performance. Tuckman labeled the stages Forming, Storming, Norming, Performing, and Adjourning, respectively. Looking back over the process of forming teams and team membership, most team members can identify these stages of development. Anticipating this process of development can help team leaders know where to facilitate forward movement and where the team process might be stuck and needs to move forward more quickly.

Salas et al. (2005) took another step into the importance of communication in research teams by focusing on "Is there a 'Big Five' in Teamwork?" The "Five" considerations include (a) team leadership; (b) mutual performance monitoring; (c) "backing-up behavior," including noticing when workloads need to be adjusted; (d) adaptability; and (e) team orientation. *Leadership* refers to the presence of leadership and direction. *Mutual performance monitoring* refers to the potential for observing the progress and performance of other team members. *Backing-up behaviors* refers to a means for monitoring the need to shift workloads or pick up help. There frequently are glitches in putting together projects, and *adaptability* is needed to make needed adjustments and retain focus. *Team orientation* refers to the sense of a team working together, or a sense of team identity. Experienced team leaders likely recognize all these issues, but those just getting involved in team leadership or team function could benefit by reading about what others have observed.

For investigators new to the grant process, learning how to work both with and within a team should be part of developing a skill set. Established investigators should maintain self-evaluations for team effectiveness to retain or improve a skill set. Poor team playing is fixable.

A review of team functioning deserves some attention given that there are few single-investigator proposals. Most proposals involve a team that plans to produce the varied components of a proposal and turn the competency of the team and its management toward a final on-time submission.

When a team is pulled together for the first time, there is a greater potential that a planned proposal will not be submitted on time or that its components will appear to be rushed. Experienced teams can respond to Funding Opportunity Announcements/Notices of Funding Opportunities (FOAs/NOFOs) that allow only a few weeks of preparation time because communications are likely better and the team is familiar with the submission process and thus proposals can be produced in a skeleton form very quickly and then filled in.

The PI's assembly of individual team members with various skills, time obligations, and deliverables requires time as well as insight into how well the proposal components are coming together. Experience is a good teacher when it comes to pulling together a team effort. Institutions with more resources can often use the skills of the grants support workforce (GSWF) to help assemble and focus the individual efforts of a team.

A team that does not submit a planned proposal on time often is a newly formed one and had to devote extra time to the assembly of the team members' individual skills, obligations, and deliverables. The review criterion tends to draw attention to individual investigator skills, but this evaluation criterion is more than that. Reviewers often comment, especially for larger projects, that the members of a team do not seem to have had much prior experience in working as a team because experienced reviewers know that communication and coordination between team members are key to managing a proposal. The more moving parts a proposal has, the more the investigators need to ensure team coordination.

Sometimes a FOA/NOFO is not spotted soon enough to pull together a reasonable proposal, or the topic showed potential for team involvement but in the end the skills did not fit the goals of the announcement. Investigators' attempts at team building that do not pan out are not necessarily wasted. If a team's submission attempts are not successful, that does not mean that potential does not exist. A team's potential should be evaluated even if a submission did not take place or was rushed and did not receive funding.

Individuals who have been involved in team management often develop rules of thumb to help facilitate better team communication. On the basis of my years of administrative experience, I offer several heuristics here. Experienced applicants/investigators who work in teams could certainly add their own rules of thumb. Ideally, evaluation by the PI should not wait until a proposal has been submitted and reviews returned; evaluation of a proposal by the principal parties should begin after the first meeting. Exhibit 11.1 provides heuristics for initial team functioning review. The list in Exhibit 11.1 has an obvious focus—communication. Communication will be a key factor throughout the development of a proposal not just for the initial meeting or the first few meetings, but if there are questions after the initial meetings about the efficacy of the items in Exhibit 11.1 then a special meeting or a survey about quality of communication might be useful. Exhibit 11.2 focuses on a presubmission productivity evaluation.

HOW COMMUNICATION IS DISTRIBUTED

A theme in this book is that grant proposals, reviews, and feedback are distributed from source to source across different platforms. Distributed cognition/communication is baked into the grant process (Elias, 2012), which means that information communicated to a team in components or pieces across different platforms and/or meetings can take on different meanings as team communication acquires a more distributed nature (Rogers & Ellis,

EXHIBIT 11.1. Suggested Team Function Evaluation Heuristics for Initial Meetings and After a Submission

1. Did all team members have a clear understanding of the goals of the FOA/NOFO?
2. Was there a clear understanding of science and personal contribution to science?
3. Was there clear communication between team members?
4. Did the meeting provide a clear means by which to communicate expected contributions?
5. Were clear timelines part of the discussion?
6. Was there a clear understanding of the resources needed (and whether, how, and where they can be acquired), including administrative and editorial support?
7. Were clear means offered by which to solve issues?
8. Were clear means offered by which team members could communicate with each other?
9. Did there seem to be buy-in from team members on goal attainment, feasibility of the process, and feasibility of the science?
10. Was a process enabled to develop trust in progression of the project and trust in leadership?

Note. FOA/NOFO = Funding Opportunity Announcement/Notice of Funding Opportunity.

EXHIBIT 11.2. Areas to Cover in a Self-, Team-, and Laboratory Production Evaluation

- Review of the literature
- Data collection
- Data analysis
- Productive laboratory or team meetings
- Publication of supportive data
- Supportive paper presentations
- On-time production
- Compliance to format and policy
- Budget development

1994; Zhang & Patel, 2006). The electronic platforms used to distribute information have structures that can influence the expression of ideas. Information transfer can become stylized or conform to convention on the basis of expectations of a format of information distribution; that is, specific language structures are expected for different kinds of communications. Academic journal submissions require a stylized presentation. Forms used to submit grant proposals tend to stylize the language and dictate how much information can be provided.

Knowledge is often vertically distributed (i.e., passed on from top to bottom) from one source to another such that the most prominent ideas make it through the process relatively intact, but the details can be lost or changed according to the personal knowledge and biases of the receiver and interpreter. Aims drift was explained in Chapter 5 as the unconscious transformation of the basic aims. Investigators working in teams should be aware that their initial ideas are often altered once the proposal process begins, so it is essential that developing ideas and nuances that are critical to the final product are communicated as a *horizontal* network aggregate to the team members rather than being distributed individually in an ipsative/individual fashion. Initial meetings are often horizontal/aggregate communication situations, but subsequent meetings are usually conducted in smaller group aggregates and from person to person, to save time. Similar to reducing aims drift, it may be useful from time to time during a proposal development to distribute information in a more horizontal and aggregate manner to reduce team cognitive drift.

Proposal communication can be a hard issue to tackle, not because of the clarity of the functions listed in Exhibit 11.1 but because individual team members' understanding and communication often moves toward personal understanding and goals. It is frequently said that there is no "I" in "team," a saying that is meant to convey that teams should work together and put aside individual concerns, but that is not plausible, especially in an academic setting. Perhaps "team" should more realistically be spelled "teIam."

Anyone who has ever attended a grant proposal meeting where their own participation is part of the eventual production process is familiar with the "I" processing focus of Exhibit 11.3. It is harder to evaluate the team and "teIam" heuristics when team leaders hold electronic, not in-person, meetings. If you cannot be in the room, it is harder to evaluate the initial team buy-in. Nevertheless, when team members' interest in participation appears to be declining, it would be useful for team leadership to reach out to see what personal issues can be resolved. An article by Zucker (2012) validates the importance I place on considering the individual in a team effort and encourages recognition of the individual in team building and management.

TEAM COMMUNICATION PLANS

Given that communication is such a vital component of grant development and submission, it is worth considering how communication between team members will be managed. The goal is not to control what is communicated but to avoid a default format in which communication becomes erratic and overwhelming.

A primary source of communication distortion is *communication overload*, a fundamental component of the life of anyone who has a computer or a phone. The requirement to make too many decisions, including what emails to read and then what emails to respond to, leads to *communication decision fatigue*.

EXHIBIT 11.3. Some of the Personal "I" Heuristics Involved in Evaluating Team Proposal Participation

Personal heuristics invoked to judge the value and progress of a team proposal include the following:

1. What is in it for me (recognition, network, money, articles, data, access to technology)?
2. Do I have the time versus the potential payoff (likelihood of success)?
3. How is time being managed for this process?
4. Do I initially like this team, team process, and development process?
5. Will my weaknesses be revealed?
6. Will my strengths be appreciated?
7. Do I have conflicting goals or interests?
8. Do I believe in the viability and feasibility of the goals?
9. Do I trust the development process (different from Item 4)?
10. Do I have any self-efficacy regarding what is conveyed or written in the application?

Cal Newport (2021), in *A World Without Email: Reimagining Work in an Age of Communication Overload,* offered significant insight into how living in a digital world has transformed the transfer of information. In speaking of email, Newport noted:

> This frenetic approach to professional collaboration generates messages faster than you can keep up—you finish one response only to find that three more have arrived in the interim—and while you're home at night, or over the weekend, or on vacation, you cannot escape the awareness that missives in your inbox are piling ever thicker in your absence. (p.43)

Ways to avoid communication overload for the grant team involve deciding what work can be completed during the initial meetings that all participants need to attend. The inconvenience of having to plan and attend group meetings can be offset by the larger inconvenience of having to fill people's inboxes with individual communications that do not capture the knowledge of what everyone was provided as a group. Discussing a communication or email plan at the beginning of a project should be part of the planning process. Choosing a means by which to place an urgency of response on email communication helps the sender think through the point of communicating and whom should be communicated with. It is equally important to communicate the importance and urgency of communicating back to the email sender and the team.

One of the more annoying aspects of email is that some recipients will continue to send emails with a subject line that was useful only for a specific communication. Having to sort back through a connected trail of emails that have been forwarded or returned under a topic that was relevant several emails ago increases reading time and leads to errors in communication. It requires discipline, but being careful to use the appropriate subject in the subject line can save everyone a great deal of frustration.

Asking email recipients to acknowledge receipt by opening the email may be all that is needed for many communications. Planning a time by which emails will be read by team members and then again by team leaders may focus email communication to bursts of emails rather than requiring one to search through the dribbles. If an accidental "reply all" is a potential work-overload issue, then try to be sure to focus on returning emails to those who need the information. Check to see who is on the emails that are returned by a reader. Does the email have to go back to all persons listed on the original email? Completed work that does not require email communication can obviously be placed in some form of a secured electronic drop box. The sad observation is that the original designed efficiency of digital communication leads to communication overload, and for each proposal there should be a plan for how the communication load will be handled.

The common methods of communication to be managed and coordinated by Systems 1, 2, and 3 include regular mail, email, personal communication, phone, text, internet meetings, and in-person group meetings. For System 1, managing these communications can be a challenge if all individuals are not present, or a summary of these communications is not provided and logged by time and date. Managing multiple communications requires effort, but miscommunication can result in more work or wasted effort. Large grants with more components, or more complex grants, are particularly susceptible to "information scramble" before a submission deadline. As mentioned, an effortful but productive way to manage these communications is to place them in a secured drop box that can be accessed by those preparing the application. At least once a week, and perhaps more often as the submission deadline nears, someone working on the application should organize the correspondence and identify out-of-date or errant information.

Employing a communication manager is a bit of a luxury but may be needed for very large grants and for those with multiple sites of function. The System 2 support workforce may be able to assist with communication management for proposals. When site communication is an integral component of a funded grant, then a communication manager can be written into the budget and promoted with a clear rationale for inclusion.

GRANT WRITING AS A DISTRIBUTED-COGNITION COMMUNICATION PROCESS

Grant writing as a technological submission and feedback process was described as a distributed-cognition process in Chapter 8 and is further discussed here as part of team communications. There exists, within laboratories and between colleagues working on a grant proposal, the potential for information to be incorrectly translated from one meeting to the next or from one individual to the next. The worst-case scenario is when the basic aims of a proposal become distorted through the information management process.

Readers may recall from previous chapters and this chapter that there are ever-evolving communication structures and processes that direct the format and content of the information within a grant proposal. The pre-submission communications are reflected in the writing of a proposal, and information is then formatted for electronic distribution to several groups of individuals. This distribution process fits within the definition of *distributed cognition* (Zhang & Patel, 2006). Proposals are not communicated in real time to panel members, so the information they contain is packaged to conform to the limitations of the communication distribution systems.

Feedback from reviewers is another form of a distributed-cognition process. Reviewers are instructed to form evaluative opinions about set issues and to report these evaluative opinions in specific ways (e.g., written comments, scores), on specific forms, under specific time constraints, and in different forums, including both private and group reporting and discussion.

In line with the theory of distributed cognition, applicants' initial ideas are not just relayed but distributed across a complex network of policy-guided communication formats that not only transfer the original ideas across time but also shape the reading context (Zhang & Patel, 2006). Feedback from reviewers, review panel members, and the panel Scientific Review Officer (SRO) is shaped by the same network of policy-guided communication formats. The format for grant application submissions likely has a trimming effect on original ideas, and the feedback from a review may suggest further trimming or alteration. Any tinkering with the nature of the review process by System 3 agents changes the distributive-cognition network in unknown ways.

There are three caveats to the applicant–reviewer–SRO communication pathways just described:

1. When System 3 review policies allow for triaged review (see Chapter 8), there is an unfiltered Pathway 2 whereby there is direct feedback from a reviewer in the form of a review and accompanying scores to show evaluation relative to a scale and perhaps to other applications.

2. The triage review process creates a reviewer's-filtered Pathway 1 whereby reviewer comments are filtered through the cognitive information provided by the comments of other primary reviewers, and panel commentary.

3. Pathway 1 reviews lead to another form of cognitive distribution. The SRO interprets the panel discussion and provides a stylized Summary Statement.

In the typical review process there is no live audition of a proposal with immediate feedback and discussion. It is likely that some individuals, including applicants, SROs, POs, and peer reviewers, communicate better across a distributed system than others. Although it is not immediately clear what the impact is when Systems 1, 2, and 3 have different communication styles, there are likely impacts on the inclusion and interpretation of content.

In Chapters 2 and 3, the grant proposal was presented as a product of policy that specifies, among other things, specific grant mechanisms and announcements that relate to grant-renewable status, applicant/investigator eligibility, FOA/NOFO-permissible goals, duration and levels of funding, and salary limitations. Policy-derived FOA/NOFO communication formats determine the specified times of for submission and review as well as the forms and formats for submission. The formats determine how much time can be given to consideration of a written proposal, what can be written, what must be written, page limitations for what is written, and the rules for communicating ideas. Very few studies have examined how changes in communication policy affect the outcome of a review.

Figure 11.1 shows a distributed-cognition perspective on grant proposal communications. In the figure, the initial ideas of the investigators/ stakeholders who incorporate the knowledge provided from the literature

FIGURE 11.1. Flow of Idea Distribution for the Review Process

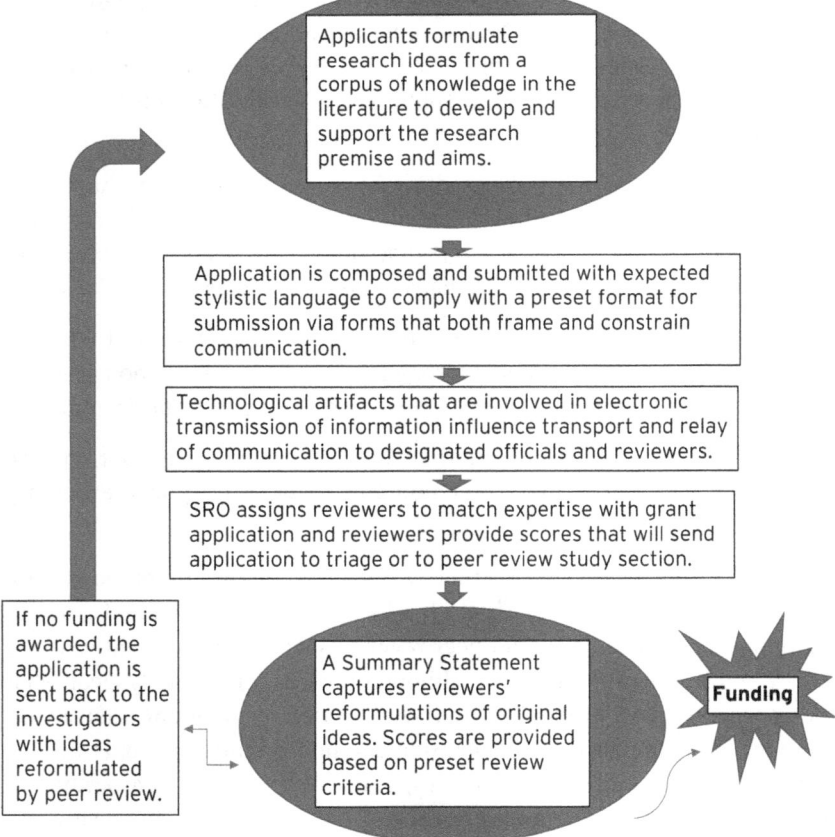

Note. SRO = Scientific Review Officer.

are submitted across a network that requires stylized communications to fit the network and external policy. This distributed-cognition product is then sent on to peer review, where information is looped back to the stakeholders and further shaped by the human artifacts of procedure and electronic transmission.

The stakeholders include the institutions that are supporting the investigators as well as the investigators themselves. A resubmission or renewal initiates a reiteration of the network information-distribution process. The scores and feedback from the Summary Statement reflect this distributed process, and the scores likely reflect not only the value of the proposed science but also the ability to communicate across the distributed network involving artifacts of policy and technology.

KEY TAKEAWAYS FROM CHAPTER 11

- One way to get a step ahead of the competition is to pay attention to time, team, and information management. With more ways to be distracted in everyday life as well as office life, developing a communication plan to eliminate extraneous information will make communicating within a team context easier and more efficient.

- Goals for communication and evaluation of the first team meeting or meetings can be established and then reviewed for success or failure.

- Application of the construct of distributed cognition to team development draws attention to how multiple communication platforms and vertical distribution of information and horizontal aggregate communications, if not monitored for content, can subtly alter initial aims and goals.

- Setting up an email plan that includes sending and response plans for individual and group emails can save all team members significant time and avoid distribution of incorrect or off-target information.

- Application due dates are not the only important dates to focus on. There are likely multiple System 2 dates for completion of forms, written documents, and signatures of signing officials. Individual team members will have dates by which to provide information. Establishing the what and when of deliverables is an important component of grant preparation, and placing an individual in charge of establishing dates for deliverables and then monitoring progress is a critical part of team management.

- There are ways for new team leaders and those new to the team process to anticipate what will occur in the team-building process. A search of the literature will reveal that there are multiple ways to examine team function. Some articles provide easy-to-understand team constructs and rules of thumb based on the experience of the author.

- Individuals function within teams, and individual perspectives and needs determine how the individual will buy into a team effort. Team leaders should not overlook individual team members' needs or characteristics.

12 BEYOND THE FIRST GRANT

Experience Provides Perspective on Writing and Funding

In Chapter 1, *grant literacy* was defined as the degree to which an individual has the capacity to obtain, communicate, process, and understand basic grant information and services in order to make appropriate grant decisions. I add to this definition that those investigators who are the most grant literate move about the world of research and grant funding with a better understanding of how to go to Plan B if Plan A does not work or is not satisfying.

An organization scheme was offered in Chapter 1 to help readers understand how the three systems of self/team (System 1), supporting institution (System 2), and funding institution (System 3) are proposed to work in an integrated fashion. Readers were offered several options for investigator/applicant self-review of System 1 functioning and perspectives on System 2 functioning.

Just one experience as an applicant/investigator who has completed a cycle of preparation, submission, review, return of the Summary Statement, and resubmission or funding increases grant literacy. One trip through the multiple System 1, System 2, and System 3 process shown in Figure 12.1 is

https://doi.org/10.1037/0000390-013
Get Funded: A Practical Guide to Understanding the Grant Application Process and Writing Winning Proposals in the Behavioral and Biomedical Fields, by J. W. Elias

FIGURE 12.1. Development of Basic Grant Literacy Components With the Potential to Develop Toward Advanced Grant Literacy Systems Integration

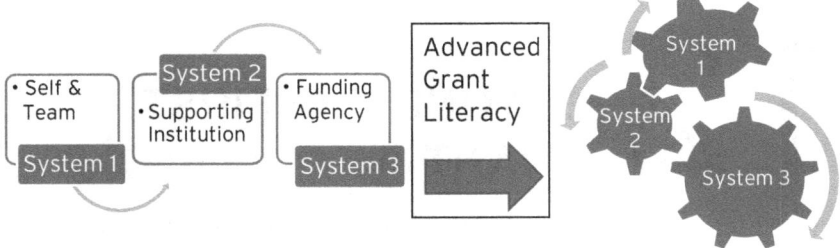

illuminating and allows investigator and team reviews to begin. An elaboration of the systems review process that was provided in Chapter 2 is repeated for the reader's convenience in Exhibit 12.1. The system review process could be applied to a specific grant proposal or a series of grant proposals.

Applicants/investigators, after three or four submissions, in particular if they were successful, should be developing a good grasp on basic grant literacy. As applicants/investigators move toward advanced grant literacy, they show evidence of better management of System 1 components and move more easily about the supporting System 2 environment to judge the degree of support needed from System 2 and System 3 to sustain the desired level of research activity, career satisfaction, and income. Figure 12.1 shows the initial straight-line knowledge of Systems 1, 2, and 3 functioning and then evolving with experience to an integrated systems knowledge illustrated as gears functioning in synchrony.

The detailed contextual components of the three proposed systems were introduced in Chapter 1 and Chapter 2 and are presented again in Exhibit 12.1.

The complexity of the three systems is apparent, and yet if an investigator is not funded on the initial submission, with enough guidance new investigators can manage their way through the systems to reapply and start through the process again. Most applications are not funded by System 3 sources on the first submission. When applications are funded, and the work begins, this experience adds another layer of skills that can be used for the next submission. New investigators funded on the first submission should take the time to look back over the functioning of System 1 and System 2 to see what was supportive and what could be changed.

Exhibit 12.1. Expanded System 1 Developmental and Management Responsibilities

1. Assimilate
   - terminology
   - communication styles
   - policies
   - forms and submission guidelines
   - opportunities and announcements
   - feedback
2. Navigate
   - the institute's organization
   - the institute's modes of communication
   - institute personnel
   - home institution
3. Communicate with
   - funding sources
   - colleagues
   - the funding team
   - the home institution
4. Accommodate
   - changing policies and rules
   - time cycles
   - positive, mixed, and negative feedback
   - changing personnel
   - advances in science
   - changing support—either more or less
5. Time management
   - time for grant development
   - time devoted to nongrant professional activities
   - personal responsibility time (nonprofessional life)
6. Develop
   - research ideas
   - research skills
   - staff
   - research team
7. Collect and publish data
   - establish data sources
   - set up data repositories
   - write and submit articles for publication
8. Motivate
   - self
   - personnel and teams
   - ideas and methods
9. Adapt to distributed cognition, that is, information distributed across and shaped by technological platforms.
10. Evaluate your satisfaction with participating in the granting process and science.

GOODNESS-OF-FIT WITH SYSTEM 2

Most researchers who work in academic positions find themselves balancing time for multiple activities, including publication, classroom teaching, mentoring, clinical activity, and administration, as well as focusing on grant funding and grant-funded projects. If possible, some prefer to focus on using grant funds to buy their time out of teaching, clinical, and/or administrative roles.

Academic positions that require one to function in a variety of human interaction roles typically entail a mixture of skills, including learning how to function in a demanding and changing social environment. In many cases, researchers require time to develop the multiple skills needed to be successful in a System 2 academic environment. New (early) investigators are learning System 2 academic skills as well as the System 1, System 2, and System 3 grant application contexts. Research training awards, such as the National Institutes of Health (NIH) K99/R00 award, fund the last 1 to 2 years of postdoctoral research and then an R01-level grant that can be used in a new academic setting. The NIH K99/R00 award does not help those new to academics learn academic skills, but it clearly helps new faculty transition to the challenge of working in an academic setting.

System 2 grant support systems must coordinate and balance their functions and resources to meet their overall System 2 mission, and that includes supporting System 1 investigators. Part of developing grant literacy for System 1 investigators and System 2 administration is understanding how grant funding and grant submission fit within the overall mission of both System 2 academic departments and System 2 as an institution.

System 2 institutions that want to support grant funding need to acquire institutional grant literacy, which includes understanding how grant literacy develops in new faculty. Researchers who acquire advanced grant literacy can, over time, develop a better understanding of how well a supporting System 2 institution can sustain an investigator with periods of low or no funding. For System 2 administrators, many of whom are also participating in System 1 submissions, knowing how well the System 1 investigator can manage the stress of dry periods of funding is part of the rationale for System 2 administration to develop advanced grant literacy. The degree to which System 2 administration can support investigators and their laboratories is a significant component of grant funding success for System 1 investigators and System 2 administration.

Peer reviews of potential and perceived environmental support from System 2 administration are scored on Summary Statements and are a significant part of sustained grant funding success for a research institution and its applicants/investigators. Top-funded System 2 supporting institutions provide strong environmental support, and strong within-institute cooperation, including training programs and research centers, to provide technical support as well as additional research funding for pilot projects. The payoff for such high-level System 2 support and cooperation is System 1's success in competing for grant funding. As discussed in Chapter 9, the expected changes in the review of NIH Research Project Grants (RPGs) are designed to reduce System 2 influences on the reviewers' scoring.

Retrospective grant submission reviews by System 2 administrators and System 1 investigators typically find that the degree of cooperation between the two systems is a key component of success. Experienced investigators and administrators can better leverage and combine System 1 and System 2 components to be more System1/System2 competitive.

When applicants/investigators move to a new System 2 environment, this results in immediate changes in the System 1 and System 2 context and culture of grant submissions and funding. A similar situation occurs for applicants/investigators when changes in existing System 2 management are implemented by System 2 management.

The System 1 and System 2 evaluation lists provided throughout this book provide a structure for evaluating or planning a move to a new research institution. If investigators who are contemplating a move to a new research environment can obtain comments on System 2 function from new environment System 1 investigators and from new environment System 2 administrators, this information can provide an integrated perspective on the new environment.

GOODNESS-OF-FIT WITH SYSTEM 3

Funding Opportunity Announcements/Notices of Funding Opportunities (FOAs/NOFOs) provide insight into the most immediate interests of a funding source, but advanced grant literacy requires knowing how personal skills and research interest fit the overall and future needs and functions within System 3 funding sources. A good way to achieve this knowledge is to go beyond simply responding to funding announcements and take advantage of descriptions of the components of a System 3 funding institution. NIH has public components of its individual institute council meetings and issues strategic goals. The NIH RePORT (https://report.nih.gov/), NIH Data Book (https://report.nih.gov/nihdatabook/), and the NIH RePORTER (https://reporter.nih.gov/) provide data on funding.

The National Science Foundation (NSF) publishes data on funding in its merit review process reports, to help educate the public and stakeholders about the breadth and scope of the funding process, and the National Science

Board (NSB) meets four times a year with sessions that are open to the public. NSF also has a visiting Program Officer (PO) system to help generate advanced grant literacy in experienced faculty.

New investigators who are working their way through the System 3 submission process may find these System 3 advanced reports less useful than more established investigators who are trying to maintain a stream of funding and who are looking for long-term investment opportunities. NSB (2020) published a document entitled *National Science Board Vision 2030*, which provides a thought-provoking look at the international expansion of science while showing how research has been influenced in the United States by foreign-born researchers. The immigration and integration process of researchers is an example of the developing international components of advanced research.

Researchers who are new to academics and who are caught up in the next funding cycle and the next FOA/NOFO might think it a waste of time to look too far into the future. For those who wish to make a long-term career out of funding in the sciences, it is worthwhile to look for vision statements from both System 2 and System 3 to see where they fit in or need to go. The National Academies of Science (Daniels & Beninson, 2018) indicated that quick action is needed to continue to train and fund the next generation of scientists; their report focuses on the transition periods into independent careers in science and offers suggestions to improve the transitional process.

Outside of the granting process, it is difficult to leverage System 3 components to gain a competitive advantage beyond using one's skills and experience to better understand and communicate with System 3. Nevertheless, knowing how to navigate and move more comfortably within the System 3 components can be an advanced grant literacy advantage.

Part of judging System 3 functioning is evaluating the quality and consistency of peer review mechanisms as they interact with the funding process. It is possible for an application to repeatedly receive peer review scores that are just outside the System 3 funding level and for the program review level to show reluctance to accept an application for funding that falls just outside the funding line. Investigators who are more advanced, especially those who have served on peer review committees, are often more willing and better able to ask the right questions to Scientific Review Officers (SROs) and POs to see what the next recommended steps might be.

If peer review scores place a submission close to a funding source's funding line, it never hurts to remind a PO that an application fits with and meets both the goals of the funding institution and the goals of the FOA/NOFO. Speaking metaphorically, some Study Sections often seem to have one foot in the past and on safe ground and the other foot in the future and ready to step on the suspension bridge over the deep gorge of new findings. Confident investigators can remind POs that the goal of the FOA/NOFO is to "build a suspension bridge across the gorge of new findings." Acquiring grant literacy helps one know where to seek leverage within each component of System 2 and System 3 and when to move on.

Attaining advanced grant literacy requires that one stay alert to changes in the context and culture of the submission and review process. Even experienced investigative teams need to determine whether their grant submission approach needs to change. Many competitive situations require that one adjust the competitive strategy or game plan while the competition is ongoing. For grant applicants/investigators, adjustments in the submission process typically begin with peer review feedback and the Summary Statement. Part of advanced grant literacy is having enough competitive events to judge how well System 1, System 2, and System 3 are working together and then making the needed adjustments to develop a stream of competitive grant proposals.

RENEWABLE GRANT OPPORTUNITIES

Both NIH and the NSF allow renewals for some grant proposals. NIH encourages it (https://www.niaid.nih.gov/grants-contracts/stayingfunded); NSF seems to allow it (https://www.nsf.gov/pubs/policydocs/pappg22_1/pappg_5.jsp#fn54). Principal Investigators (PIs) are advised that the NSF Accomplishment-Based Renewal is a special type of proposal that is appropriate only for an investigator who has made significant contributions, over several years, in research addressed by the proposal. Investigators are strongly urged to contact the appropriate NSF PO before developing a proposal using this format.

Applicants should check the initial FOA/NOFO for a funded application to see if renewal applications are possible. Smaller grant applications, such as the NIH R03s and R21s, are not renewable unless stated otherwise. A successful R21 ideally will produce data to promote a larger R01 new submission. Some R01-level grants are not renewable.

Applicants planning to renew a previously funded and now-ending grant should schedule a discussion with the PO as a first step. Interest in funding at the program level on the renewal topic of your application might be limited. If renewals are possible for a previously funded application, it may be that the field has moved forward beyond interest in the renewal or that the field has been saturated with applications on the topic of the renewal. A search of the NIH-funded grant database by topic area (https:// orip.nih.gov/funding/search-awarded-grants) will tell you how much work is currently or funded, or has previously been funded, in your renewal topic area and whether that work has resulted in publications to be reviewed or whether the funding institution was the home for many of the grants in your proposed topic area.

Both NSF and NIH funding sources suggest planning renewals in advance of the end of a current grant proposal. NSF suggests planning a renewal as though it is a new grant that contains progress from the previous grant as support for the premise. NIH allows for a submission when progress on previous aims is provided. NIH offers guidance to funded investigators (https://orip.nih.gov/funding/search-awarded-grants) to help develop progress reports that will indicate if further development in the form of renewal seems likely. Previous Summary Statement reviews may be provided to the peer review Study Section. If there are strong progress reports, including publications, and the progress suggests further exploration of the initial aims, then applying before the end date of the initial application is a good strategy.

If it is not clear how the previous aims should be pursued with a renewal, many established investigators will indicate that using a little time to gain perspective on outcomes from the ending grant is a good strategy. Sometimes it is an ancillary study supported by a grant that produces the next best opportunity, and further pursuit of older aims may not be as productive in the long term as pursuing a new direction. Not all outcomes of grant proposals, even if important, mean continuing in the same direction. The feeling of being rushed on a renewal to keep the funds flowing may be a sign that a renewal is not the best approach to continuing an immediate stream of funding.

DEVELOPING A CONTINUAL STREAM OF FUNDING

If an agency such as NIH makes an investment in an investigator by funding a research grant, then encouraging a future string of research projects from that investigator makes sense. Reports from NIH seem to suggest that an NIH and perhaps investigator goal of a stream of funding is achieved when investigators receive sequentially funded R01 grants or R01-equivalent grants with little break in funding. NIH defines a *stream of funding* as continual funding from NIH. The definition does not include maintaining a stream of funding by using multiple sources of funding. This is a bit of a self-interest perspective from NIH, which, as stated, naturally would like investments in investigators to promote more grants. Ironically, NIH has also provided data showing how difficult it is to maintain continuous NIH R01 funding (Lauer, 2021; Lauer & Collins, 2018).

Based on NIH data (Lauer, 2021), the PI age for receiving an initial funded grant application had a mean of 40 years (*Mdn* = 38) in 1995 compared with an increased mean of 44 years (*Mdn* = 42) in 2020. In 2020, the PI first-time funding age for the 90th percentile was 54. The data on age at first funding revealed no differences between men and women. Despite the increasing age for first-time funding for PIs from 1995 to 2020, the number of first-time–funded PIs in 2020 (1,342) was roughly 50% more than in 1995 (Lauer, 2021).

Lauer (2020) provided a discussion of the probabilities of maintaining a stream of funding at the RPG level. On Lauer's page, readers will find tables that are broken down into the defined categories of (a) Early-Stage Investigators (ESIs), (b) New Investigators (beyond the defined ESI limits), (c) At-Risk investigators (funded but not re-funded within a year or two or more), and (d) Established Investigators (who maintain a stream of funding). The NIH blog *Open Mike* (https://nexus.od.nih.gov/all/category/blog/) frequently discusses funding context issues related to competition and grant writing for early and new investigators.

The online accompanying blog commentary from NIH (*Extramural Nexus*, https://nexus.od.nih.gov/all/), which follows up on streams of funding data, is highly informative with respect to the various perspectives on why it takes so long to receive initial funding. The time from first to second R01 funding as a PI is not published by NIH, but the potential impact receiving funding late in one's career could be followed by a long period until the next NIH-funded application as a PI.

Part of the delay in funding for a second NIH grant is a delay in R01 resubmissions after the first funded R01 (Antman et al., 2021). More NIH second-funding applications are being submitted as new applications by the newer cohorts of investigators. That may not be a bad strategy. It may be a mistake to quickly organize a renewal application without carefully considering the data outcomes of a grant that is just ending or has ended. Perspective and reflection are powerful tools when reviewing the progress of a recent grant. It can take some time to uncover the real value in data. A diversified funding approach can help provide the time to mull over hypothesis-driven data from the last funded application.

In addition to data provided by NIH on the *Extramural Nexus* blog, the National Cancer Institute (NCI) published an analysis of age to first NCI grant as a new PI (Antman et al., 2021) that corroborates concerns about the aging of the NIH workforce and used data from 1990 to 2016. The trepidation expressed in the article is not that careers are extended by later initial PI funding status, although career extension might be a consequence of later-in-career initial funding, but that the time from essential training to PI status is being extended. Equally worrisome to NCI is the fact that the data from the NCI 1990-to-2016 analysis show that there has been a significant drop after initial funding in resubmissions and in following up the initial R01 with a stream of funding. The data reported do not include funding outside of NIH or NIH funding from non-R01 mechanisms.

NSF has not provided similar kinds of data regarding the age of the investigator or time to PI status and age at first funding. Time to first funding is not a major point of discussion on the NSF website. NSF is a major contributor to concerns about the STEM training of the national scientific workforce and is involved in the effort to reduce disparities in training and funding (Mervis, 2022). It would be useful to know whether the age trends for two of the major sources of science funding in the United States see similar age trends at the earliest ages of funding.

REDEFINING A STREAM OF FUNDING

Regardless of the NIH's perspective, however, the investigator's individual circumstances determine what is enough funding within a particular time period. Investigators and supporting institutions can easily define "a flow of funding" as receiving support from a variety of agencies, including support from a professional society. Thinking and planning in terms of multiple sources of funding is advisable if the goal is to never miss a pay period without funding. This typically requires a variety of research interests and the ability to submit applications in multiple areas with aims that do not overlap. A skill set and a mindset that allows for collaboration as a co-investigator on grants can produce a stream of individual support and publications even if the PI's investigator status is intermittent.

An investigator's goal may be to maintain or cover a gap in salary for a certain period while still achieving multiple career goals such as publication/citation, supporting/training students, editing/reviewing, participation in professional society activities, and serving in a System 2 administrative role. Academic positions often require participation in multiple roles to achieve tenure and/or promotion or salary increases. Performance in multiple roles can be challenging if there is a requirement for faculty to serve in permanent

PI status. The early investigator percentile advantage in funding from NIH institutes helps with the initial tenure status issue. Maintaining continual funding can be a challenge if researchers apply their grant literacy to only one funding source.

INEQUITIES IN FUNDING: RACIAL CONCERNS

Several articles published by either NIH or that are based on NIH data have examined inequities in funding. Chen et al. (2022) used NSF Merit Review Digest reports to examine data from 1999 to 2019 to report on approximately 50,000 NSF applicants who submitted applications. The analysis showed that there were racial inequities in NSF funding, such that Asian, Black, and Hispanic/Latino applicants received fewer funded grants in total and by the percentage of applications. This report acknowledged the issues of trying to define race as a construct, race as an identity, racial identification, and ethnic status. There is a trend toward applicants not identifying themselves by racial category or sex/gender categories on applications.

The lack of identification and checks on the authenticity of identification makes finding usable data for analysis more difficult. Nevertheless, there are multiple reports of inequities in funding by age, gender, and racial category (Antman et al., 2021; Chen et al., 2022; Conte et al., 2020; Eblen et al., 2016; Lauer, 2022a; Lindner et al., 2016; Pickett, 2018).

Chen et al.'s (2022) discussion of NSF funding inequities cites many of the articles, editorials, blogs, and perspectives offered in the literature on racial disparities in NIH grant funding. The racial inequities in funding are the most socially distressing and most likely to be termed systemic, therefore mimicking current societal issues.

Funding agencies and peer review sections both struggle with the overlapping definitions of race and ethnicity relative to self-identity and other-identity. Definitions of race and identity and their utility are being challenged in the literature (Bryce, 2023). It is hard for reviewers to completely ignore race as an issue in reviewing grant applications because, as discussed in Chapter 8, the older definitions of race and ethnicity of potential subjects are still placed into categories for inclusion coding at the point of peer review (see codes provided for human subjects, minority status, gender, and age at https://www.niaid.nih.gov/grants-contracts/human-subjects-inclusion-codes).

A common point of discussion in Study Sections is that, because of regional demographics, an application might satisfy minority enrollment issues for two racial categories but not a third. The reviewers can provide a note to the applicants/investigators in the review comments or invoke a code that indicates a bar to funding. The code can be appealed to and removed only by the System 3 funding agency. The assignment of a code is a judgment, as is the expunging of a code. Enrollment issues are important ones in encouraging the development of a research community that includes representative samples of the population.

INEQUITIES IN FUNDING: AGE CONCERNS

Daniels (2015) provided an extensive discussion of how the experience of older applicants leads to higher funding percentages in the older age groups. Exhibit 12.2 shows the potential advantages for previously funded applicants. The term *potential* is used in the exhibit because many investigators who are new to funding do not have a full understanding of the conditions that led to initial funding. As suggested in Chapter 8, a failure to immediately evaluate the production qualities of a funded application may result in a loss of valuable information to apply to future submissions.

In 2017, NIH addressed funding issues for new investigators relative to the success of older and more grant-literate investigators by proposing a limit on the number of grants a PI could receive based on a point system assigned to each mechanism/Activity Code (Lauer, 2017). Feedback from the established research community and the sheer difficulties of maintaining a point system in such a fluid environment as grant funding resulted in an abrupt change of plans. The change in the NIH plan emphasized providing funds

EXHIBIT 12.2. The Critical Elements of Grant Competition That Should Favor Previously Funded Applicants

A potentially better understanding of the following elements should help applicants who have been previously funded:

- how to shape the Aims, Significance, Innovation, and Approach sections to make a convincing argument
- how to follow policy changes that affect grant writing
- research team function and research team management, including research institution personnel
- the need for early data acquisition/analysis in a grant to support publications and provide pilot data for the next application
- data acquisition and analysis
- time allotted for submission and time management before and after starting the submission process
- how to obtain and retain resources, including personnel, laboratory resources, and a better ability to set aside time to execute a funded grant and plan for another

to new investigators by establishing a different review process (clustered by status) for early and new investigators (see https://www.niddk.nih.gov/ research-funding/process/apply/new-early-stage-investigators) and a more advantageous funding line. That change seems to have made a significant difference in funding for investigators who are early in their careers. However, once an investigator has been funded for an NIH R01 or equivalent grant, the early status and new investigator status no longer apply.

Even with the loss of early/new investigator status, with a funding credential many investigators will have an opportunity to participate in peer review as a Study Section participant or member. This is an opportunity to evaluate the review process firsthand. Investigators sitting on review panels can improve their grant literacy and see how other scientists form their aims and other sections of the proposal and how resubmitted grants use the 1-page explanation to answer previous review concerns. Investigator participation on a peer review panel cannot fully reveal the intricacies of the process, but it does allow the opportunity to see how good management of all the components of a proposal can result in better scores. Participating on a peer review panel allows the reviewer to clearly understand how published data from previous or ongoing funded research provides an advantage for supporting the premise or the methods of an application.

ADDITIONAL FACTORS RELATED TO INEQUITIES IN FUNDING

Given the overall funding rate of science proposals there is inequity in the range of scores of applications that are funded. Elias and Elias (2012) commented on how low funding for research can potentially affect reviewer scoring because reviewers know higher scores will be problematic. The percentiles of scoring that represent success and failure can become compressed toward the better scoring end. When scores become compressed, small differences in scoring become more important. Greater funding to some institutions and not others has been identified as a source of inequity by the Center for Scientific Review (CSR; see Chapter 8), and the CSR's proposed changes in the review are focused on how applications can be reviewed without knowing what the institution is. This will be a difficult task for CSR, because whether a similar condition will be made for program funding is not clear.

The multiple investigations by NIH staff and investigators outside of NIH using NIH data suggest that differences in funding by sex have been reduced over the years, with women moving toward becoming more equal to men in terms of funding status. Defining "more equal" in terms of 20 years of data,

Safdar et al. (2021) examined data extracted from NIH RePORTER for the years 1999–2019. Research grants funded included extramural awards given to research centers, research projects, SBIR/STTR awards, and other research grants. The overall increase in funded grants submitted by females from 1999 to 2019 ranged from 26% to 34%, for an estimated increase of 11% in funding. Males showed roughly 6% to 7% in overall funding. For women, the specific increase in the percentage of RPGs as a separate category was 8%. RPG funding increased from 26% in 1999 to 34% in 2019. The Research Career Awards increased from 35% to 53%. Although progress may seem slow, funding for women has been steadily increasing.

One trend is clear: Repeated (not intermittent) success in receiving funded NIH grants provides an NIH grant submission advantage. All System 1 investigators, and all System 2 institutions, are not equal with respect to the odds of receiving a peer review score that will lead to funding, or even encouragement to resubmit. Public and private commentary on inequity issues often adheres to Miles's Law, "Where you stand is where you sit" (Miles, 1978).

FUNDING INEQUITIES ARE NOT ABSOLUTE OR WELL EXPLAINED

Funding inequities by class are not absolute in the sense that certain classes or groups never receive funding and other classes and groups always receive funding. Examination of the data supplied by the articles cited in this chapter suggests that there are overlapping distributions of funding between the classes represented. Examining the roles and skills identified for System 1, System 2, and System 3 makes the source of inequities related to a particular part of a complex distributed communication process hard to pinpoint.

There may be multiple points in the grant preparation process at which there is greater potential for inequity in application preparation and that results in poorer scoring applications. Potential sources of inequities in application preparation could be better explored and might include one or more of the following:

- training opportunities;
- pilot data and publication opportunities;
- greater or lesser access to System 2 grant preparation support;
- reduced access to resources, including time, environmental support, and access to colleagues and collaborators;
- literature available to support a premise that is just developing or contradictory;

- ability to meet the funding institute's interests;
- the topic of interest as it relates to the number of applicants competing for funding; and
- funding opportunities for topics of interest.

For the last point (topic of interest), Hoppe et al. (2019) found that topic choice accounted for 20% of the variance in funding after several other factors were taken into account.

Because System 3 funding institutions provide peer review and funding, this would be the natural starting point for an examination of processes that lead to inequities. Therefore, the issue of inequity in funding is both personal and social. Reviewer bias may be a point of conversion to inequities in scoring or that may be the simplest reified explanation (description without complete explanation) for a more complex process that eventually reveals itself as inequities in funding.

THE VALUE OF READING ABOUT INEQUITIES

It is important to address, on a national level, the sources of inequities, to ensure that there are not induced reductions in submissions that lead to inequities of funding. The value of reading articles and listening to discussions about inequities is not to be discouraged but to see how to better position oneself with respect to developing a science career that can provide financial support and self-efficacy. Part of learning the culture of grant funding is recognizing that there will be inequities relative to resources. It is hard to eliminate the idea of advantage in such a complex context as grant funding, where competition is promoted and opportunities are made available to gain advantages (e.g., training awards).

A POTENTIAL SOLUTION TO NEW-FACULTY INEQUITIES IN ACADEMICS AND GRANT FUNDING

New academicians must acquire the skills to manage multiple roles. If achieving funding is one of those roles, then adjustment to academics competes with the development of grant literacy. Figure 12.1 depicts the development of grant literacy, which results in a greater understanding of the integration of System 1, System 2, System 3 relative to grant funding. Not depicted in the figure are any experiences that lead to a better integration of all the components of a faculty position.

A potential solution to diversity, sex/gender, and age inequities in grant funding is to consider the difficulty that new faculty have in adapting to new requirements and new roles. System 2 administrations could help with the acquisition of new skills for new roles by providing reduced academic loads and access to integration-into-academics mentoring as part of an entry-level plan to support grant proposals.

The purpose of the proposed support would be to help new faculty learn how to manage the multifunctional academic role, which would include reduced coursework loads and mentoring that focuses not only on how to achieve grant funding (which by itself could add to the complexity of a new academic role) but also on adjusting to the multiple challenges of an academic environment.

System 3 funding sources could develop block grants for academic support to System 2 institutions that would be distributed to new System 1 academicians. STEM training prepares one for the development of science, but not necessarily for personal development within research and/or academic roles. Mentor supplements in terms of a reduced course load or salary would be part of the block grants to System 2 institutions.

System 2 institutions would not have to be dependent on System 3 support; they could develop their own within-institution block grants to schools, colleges, and departments. The length and degree of support would need to be determined by the System 2 institutions. A question that has not been asked of diverse grant applicants is, "What role does general job satisfaction play in your ability to prepare grants or plan for a submission when the initial application does not receive funding?"

WHERE IS RESEARCH FUNDING HEADED?

The Consolidated Appropriations Act of 2022 (Pub. L. 117-103; https://www.congress.gov/117/plaws/publ103/PLAW-117publ103.pdf) was enacted on March 15, 2022, and authorized $1 billion in funding to establish the Advanced Research Projects Agency for Health (ARPA-H) within the U.S. Department of Health and Human Services. The Congressional Budget Office provides an overview of the legislation at https://sgp.fas.org/crs/misc/R47074.pdf. The monies and new institute were established to fund health-related grants that support projects involving more risk and fewer pilot data, which is a mission similar to the military's Defense Advanced Research Projects Agency (DARPA; https://www.darpa.mil).

Obviously, there was a desire at the time of the bill passing to move health-related science forward at a more rapid pace that could not be met in the standing study sections at NIH (Mesa, 2022). The suggestion to place the new institute at NIH but require direct oversight by the Department of Health and Human Services represents a desire to avoid the more standard review processes. It is not clear at the time of this writing whether the federal research funding agencies will be reduced in budget and staff or what effect such proposed reductions would have on the new agency or the existing and operative federal research agencies.

An additional major concern for federal research funding agencies is that the current NIH and NSF funding patterns reveal inequities in funding such that women and/or racial and ethnic minorities receive less funding and fewer funded grants. The current approach to addressing the inequities is to focus on changing the review process to initially reduce the impact of the applicant/investigator environment and prior investigator success on the initial peer review ratings. Although Study Sections at NIH do not discuss funding, the implication is that a difference in scoring would permit more minority applicants and female applicants to be reviewed at the program funding level with scores that are within the institute's funding lines.

Thus, as we head into the middle of the 2020–2029 decade, System 1, System 2, and System 3 participants in the granting process will try to manage changes in the review process as well as support funding goals that will maintain established grantees, renew funding for recent grantees, and bring into the systems new/early funded grantees. Budgeting will play a significant role in how well the changes in review are implemented and how well researcher–workforce goals are obtained.

KEY TAKEAWAYS FROM CHAPTER 12

- Successful grant writing requires that one put in the hours to learn the language and the culture while developing the ability to adjust and adapt to the months-long cycles of submission and review.

- Once learned, the language of grant funding tends to change slowly, but the opportunities to advance science via grant funding can change quickly. For "early," "new," and "advanced" applicants there is always something new to learn and apply with each grant submission.

- FOAs/NOFOs induce the changes needed to meet the goals and the agendas of the funding agencies, including the ethical treatment of human and animal subjects, the ethical reporting of data, and standards for sharing data collected with the help of public dollars.

- Funding institutes try to set budgetary interests for several years for planning purposes, but funding is done on a yearly basis and the possibility of receiving funding can improve or decline over the multiple cycles in each budget year.

- Applicants/investigators who wish to maintain a stream of funding should pay attention to the multiple sources for funding and posted funding lines.

- The following are advanced grant literacy heuristics:
  - Time, as managed by System 1 applicants/investigators and allotted by System 2 administrators, is critical to data access and analysis and grant submission.
  - There will be dry spots for funding for even the most experienced investigators and their teams.
  - There will be changes in plans and support based on the comings and goings of System 2 and System 3 administrators.
  - There will be limited funding for some interest areas as other areas gain interest.
  - Maintaining a stream of funding requires consistent resources, shown in Exhibit 12.1.
  - Team management and team communication plans are becoming more important to the ability to manage the distribution of complex ideas and plans.
  - Science pursued through research is resource dependent.
  - Experienced investigators tend to know what to do next, whether in response to receiving funding for a project, planning for a project in response to funding coming to an end, or by diversifying a funding portfolio.

References

Aliouche, H. (2022, July 15). *What is reproducibility?* https://www.news-medical.net/life-sciences/What-is-Reproducibility.aspx

Antman, M. D., Gorelik, R., Kennedy, A., Liou, G. F., Billingslea, E. N., Corrigan, J. G., & Bennett, L. M. (2021). Changes in the National Cancer Institute's R01 workforce: Growth, aging, retention, and policy implications. *The Journal of Clinical Investigation, 131*(7), e146925. https://doi.org/10.1172/JCI146925

Bayh–Dole Act of 1980, Pub. L. 96-517, 94 Stat. 3015, and in 35 U.S.C. § 200–212.

Bourgeois, D. T. (2014). *Information systems for business and beyond.* Lulu.com.

Bryce, E. (with Pappas, S.). (2023, February 1). What's the difference between race and ethnicity? *Live Science.* https://www.livescience.com/difference-between-race-ethnicity.html

Byrnes, N. (2022, December 8). *Proposed new framework for NIH peer review criteria.* https://www.acd.od.nih.gov/documents/presentations/12082022_Proposed_Changes_to_Peer_Review.pdf

Byrnes, N., & Lauer, M. (2022, December 8). Update on simplifying review criteria: A request for information. *Review Matters.* https://www.csr.nih.gov/reviewmatters/2022/12/08/update-on-simplifying-review-criteria-a-request-for-information/

Byrnes, N., & Lauer, M. (2023, April 25). Update on improving fellowship review: A request for information. *Review Matters.* https://www.csr.nih.gov/reviewmatters/2023/04/25/update-on-improving-fellowship-review-a-request-for-information/

Chen, C. Y., Kahanamoku, S. S., Tripati, A., Alegado, R. A., Morris, V. R., Andrade, K., & Hosbey, J. (2022). Meta-research: Systemic racial disparities in funding rates at the National Science Foundation. *eLife, 11*, e83071. https://doi.org/10.7554/eLife.83071

Consolidated Appropriations Act of 2022, Pub. L. 117-103, 136 Stat. 49. https://www.congress.gov/117/plaws/publ103/PLAW-117publ103.pdf

Conte, M. L., Schnell, S., Ettinger, A. S., & Omary, M. B. (2020). Trends in NIH–supported career development funding: Implications for institutions, trainees, and the future research workforce. *JCI Insight, 5*(17), e14817. https://doi.org/10.1172/jci.insight.142817

Daniels, R. J. (2015). A generation at risk: Young investigators and the future of the biomedical workforce. *Proceedings of the National Academy of Sciences, 112*(2), 313–318. https://doi.org/10.1073/pnas.1418761112

Daniels, R. J., & Beninson, L. (Eds.). (2018). *The next generation of biomedical and behavioral sciences researchers: Breaking through*. National Academies Press.

Distributed cognition. (2023, March 26). In *Wikipedia*. https://en.wikipedia.org/wiki/Distributed_cognition

Eblen, M. K., Wagner, R. M., Chowdhury, D. R., Patel, K. C., & Pearson, K. (2016). How criterion scores predict the overall impact score and funding outcomes for National Institutes of Health peer-reviewed applications. *PLOS ONE, 11*(6), e0155060. https://doi.org/10.1371/journal.pone.0155060

Elias, J. W. (2009). *Improving health literacy for older adults: Expert panel report 2009*. U.S. Department of Health and Human Services.

Elias, J. W. (2012, November 4–18). *Grant submission and review: Application of cognitive aging and distributive cognition* [Paper presentation]. "Charting New Frontiers in Aging": 65th Annual Meeting of the Gerontological Society of America, San Diego, CA, United States.

Elias, J. W., & Elias, M. F. (2012). Low funding has potential to start United States National Institutes of Health peer review down a path of unintended consequences. *Experimental Aging Research, 38*(4), 460–467. https://doi.org/10.1080/0361073X.2012.699814

Ferreira, F. (2021). In defense of the passive voice. *American Psychologist, 76*(1), 145–153. https://doi.org/10.1037/amp0000620

Food and Drug Administration Modernization Act of 1997, Pub. L. 105-115, 11 Stat. 2310. https://www.govinfo.gov/content/pkg/PLAW-105publ115/pdf/PLAW-105publ115.pdf#page=16

Forms Status Report. (n.d.). https://www.grants.gov/web/grants/forms/forms-status-report.html

Gill, J. (2022, September 29). *Congress saves SBIR program at the last minute, with strings attached*. Breaking Defense. https://breakingdefense.com/2022/09/congress-saves-sbir-program-at-the-last-minute-with-strings-attached/

Hoppe, T. A., Litovitz, A., Willis, K. A., Meseroll, R. A., Perkins, M. J., Hutchins, B. I., Davis, A. F., Lauer, M. S., Valantine, H. A., Anderson, J. M., & Santangelo, G. M. (2019). Topic choice contributes to the lower rate of NIH awards to African-American/black scientists. *Science Advances, 5*(10), eaaw7238. https://doi.org/10.1126/sciadv.aaw7238

Kahneman, D., & Miller, D. T. (1986). Norm theory: Comparing reality to its alternatives. *Psychological Review, 93*(2), 136–153. https://doi.org/10.1037/0033-295X.93.2.136

Landrum, R. E. (2021). *Undergraduate writing in psychology: Learning to tell the scientific story* (3rd ed.). American Psychological Association. https://doi.org/10.1037/0000206-000

Lauer, M. (2016, January 28). Scientific premise in NIH grant applications. *Open Mike.* https://nexus.od.nih.gov/all/2016/01/28/scientific-premise-in-nih-grant-applications/

Lauer, M. (2017, May 2). Implementing limits on grant support to strengthen the biomedical research workforce. *Extramural Nexus.* https://nexus.od.nih.gov/all/2017/05/02/nih-grant-support-index/

Lauer, M. (2019, October 10). Delving further into the funding gap between White and Black researchers. *Open Mike.* https://nexus.od.nih.gov/all/2019/10/10/delving-further-into-the-funding-gap-between-white-and-black-researchers/

Lauer, M. (2020, February 7). What's happening with "at-risk" investigators? *Open Mike.* https://nexus.od.nih.gov/all/2020/02/07/whats-happening-with-at-risk-investigators/

Lauer, M. (2021, November 18). Long-term trends in the age of principal investigators supported for the first time on NIH R01-equivalent awards. *Extramural Nexus.* https://nexus.od.nih.gov/all/2020/02/07/whats-happening-with-at-risk-investigators/#:~:text=In%20summary%2C%20like%20with%20early,least%20one%20more%20fiscal%20year

Lauer, M. (2022a, January 18). Inequalities in the distribution of National Institutes of Health research project grant funding. *Extramural Nexus.* https://nexus.od.nih.gov/all/2022/01/18/inequalities-in-the-distribution-of-national-institutes-of-health-research-project-grant-funding/

Lauer, M. (2022b, April 5). Introducing NIH's new scientific data sharing website. *Extramural Nexus.* https://nexus.od.nih.gov/all/2022/04/05/introducing-nihs-new-scientific-data-sharing-website/

Lauer, M., & Bernard, M. A. (2022, June 14). Research project grant funding rates and principal investigator race and ethnicity. *Extramural Nexus.* https://nexus.od.nih.gov/all/2022/06/14/research-project-grant-funding-rates-and-principal-investigator-race-and-ethnicity/

Lauer, M., & Byrnes, N. (2023, April 25). Update on improving fellowship review: A request for information. *Extramural Nexus.* https://nexus.od.nih.gov/all/2023/04/25/update-on-improving-fellowship-review-a-request-for-information/

Lauer, M., & Collins, F. (2018, May 4). The issue that keeps us awake at night. *Extramural Nexus.* https://nexus.od.nih.gov/all/2018/05/04/the-issue-that-keeps-us-awake-at-night/

Lindner, M. D., Vancea, A., Chen, M.-C., & Chacko, G. (2016). NIH peer review: Scored review criteria and overall impact. *American Journal of Evaluation, 37*(2), 238–249. https://doi.org/10.1177/1098214015582049

Mervis, J. (2022, January 21). U.S. science no longer leads the world. Here's how top advisers say the nation should respond. *Science Insider.* https://doi.org/10.1126/science.ada0431

Mesa, N. (2022, April 1). *ARPA-H to be within NIH but independently managed by HHS*. https://www.the-scientist.com/news-opinion/arpa-h-to-be-within-nih-but-independently-managed-by-hhs-69862

Miles, R. E. (1978). The origin and meaning of Miles' Law. *Public Administration Review, 38*(5), 399–403. https://doi.org/10.2307/975497

National Defense Authorization Act for Fiscal Year 2017. Pub. L. 114-328, 130 Stat. 2000. https://www.congress.gov/114/plaws/publ328/PLAW-114publ328.pdf

National Institutes of Health. (n.d.). *What we do: Budget*. Retrieved January 20, 2023, from https://www.nih.gov/about-nih/what-we-do/budget

National Institutes of Health. (2014, October 23). *Notice of revised NIH definition of "clinical trial"* (Notice NOT-OD-15-015). https://grants.nih.gov/grants/guide/notice-files/NOT-OD-15-015.html

National Institutes of Health. (2016, September 16). *NIH Policy on the dissemination of NIH-funded clinical trial information* (Notice NOT-OD-16-149). https://grants.nih.gov/grants/guide/notice-files/NOT-OD-16-149.html

National Institutes of Health. (2017, December 19). *Revision: NIH policy and guidelines on the inclusion of individuals across the lifespan as participants in research involving human subjects* (Notice NOT-OD-18-116). https://grants.nih.gov/grants/guide/notice-files/NOT-OD-18-116.html

National Institutes of Health. (2019a, January 7). *Comparison of funding opportunity announcement types by clinical trial allowability*.

National Institutes of Health. (2019b, March 18). *Guidelines for the review of inclusion on the basis of sex/gender, race, ethnicity, and age in clinical research*. https://grants.nih.gov/grants/peer/guidelines_general/Review_Human_Subjects_Inclusion.pdf

National Institutes of Health. (2021, March 5). *Review criteria at a glance*. https://grants.nih.gov/grants/peer/guidelines_general/Review_Criteria_at_a_glance.pdf

National Institutes of Health. (2022, August 30). *Continued extension of certain flexibilities for prospective basic experimental studies with human participants* (Notice NOT-OD-22-205). https://grants.nih.gov/grants/guide/notice-files/NOT-OD-22-205.html

National Institutes of Health, National Institute of Allergy and Infectious Diseases. (2021, June 16). *New formats and instructions for other support and biographical sketches*. https://www.niaid.nih.gov/grants-contracts/other-support-and-biosketches

National Science Board. (2020, May). *Vision 2030* (Report No. NSB-2020-15). https://www.nsf.gov/nsb/publications/2020/nsb202015.pdf

Newport, C. (2021). *A world without email: Reimagining work in an age of communication overload*. Portfolio.

Office of Disease Prevention and Promotion. (2019). *Healthy People 2030: Health literacy definition public comments*. https://health.gov/sites/default/files/2020-08/HP2030_Health-Literacy-Definition-Public-Comments_508.pdf

Patient Protection and Affordable Care Act, Pub. L. 111-148, 42 U.S.C. §§ 18001–18121 (2010).

Pickett, C. (2018, July 19). Examining the distribution of K99/R00 wards by race. *BioSciPol.* https://bioscipol.blogspot.com/2020/12/examining-distribution-of-k99r00-awards.html

Pickett, C. L. (2019). The increasing importance of fellowships and career development awards in the careers of early-stage biomedical academic researchers. *PLOS ONE, 14*(10), e0223876. https://doi.org/10.1371/journal.pone.0223876

Radley, G. (2019, May 10). Recognize and make nominalizations work for you. *The Writer.* https://www.writermag.com/improve-your-writing/revision-grammar/nominalizations/

Reed, B. (2020, November 19). Should we keep meeting this way? *Open Mike.* https://nexus.od.nih.gov/all/2020/11/19/should-we-keep-meeting-this-way/

Rogers, Y., & Ellis, J. (1994). Distributed cognition: An alternative framework for analysing and explaining collaborative working. *Journal of Information Technology, 9*(2), 119–128. https://doi.org/10.1177/026839629400900203

Safdar, B., Naveed, S., Chaudhary, A. M. D., Saboor, S., Zeshan, M., & Khosa, F. (2021). Gender disparity in grants and awards at the National Institute of Health. *Cureus, 13*(4), e14644. https://doi.org/10.7759/cureus.14644

Salas, E., Sims, D. E., & Burke, C. S. (2005). Is there a "Big Five" in teamwork? *Small Group Research, 36*(5), 555–599. https://doi.org/10.1177/1046496405277134

SBIR and STTR Extension Act of 2022, Pub. L. 117-183, S 4900 117th. https://www.congress.gov/bill/117th-congress/senate-bill/4900

Small Business Innovation Development Act of 1982, Pub. L. 97-219, 96 Stat. 217. https://www.congress.gov/bill/97th-congress/senate-bill/881/text

Tuckman, B. W. (1965). Developmental sequence in small groups. *Psychological Bulletin, 63*(6), 384–399. https://doi.org/10.1037/h0022100

United States Innovation and Competition Act of 2021, Pub. L. 117-167, H.R. 4346, 117th Congress. https://www.govinfo.gov/app/details/COMPS-11490

U.S. Department of Health and Human Services, Office for Human Research Protections. (n.d.). *Federal policy for the protection of human subjects ("common rule").* https://www.hhs.gov/ohrp/regulations-and-policy/regulations/common-rule/index.html

What Did the Bayh–Dole Act of 1980 Do? (2022). Federal Prism. https://federalprism.com/what-did-the-bayh-dole-act-of-1980-do/

Young, D. W. (Director). (2019). *The Booksellers* [Film]. Blackletter Films. https://www.imdb.com/title/tt9355194/

Zhang, J., & Patel, V. (2006). Distributed cognition, representation, and affordance. *Pragmatics and Cognition, 14*(2), 333–341. https://www.doi.org/10.1075/pc.14.2.12zha

Zucker, D. (2012). Developing your career in an age of team science. *Journal of Investigative Medicine, 60*(5), 779–784. https://doi.org/10.2310/JIM.0b013e3182508317

Index

About the Author

Jeffrey Wayne Elias, PhD, had a precollege interest in the design of research studies which led to a progression of college, graduate school, and post-graduate research studies focused on statistical design and analysis. Research grants and publications resulted from these early interests. The focused interest in the design of research led to 27 years as the editor-in-chief of the journal *Experimental Aging Research* and multiple recognitions of research accomplishment. These awards include the John Gilmour Prize for Psychology from Allegheny College, the Public Health Service Fellowship from West Virginia University, the Texas Tech University Arts and Sciences Research Award, the National Institute on Aging Director's Award (shared) for developing the Neuroeconomics Initiative, and the National Institute on Aging Director's Award for the Neuroepidemiology Team (architects of the NIH Toolbox). Fellow status is held in the Gerontological Society, the American Psychological Association, and the Association for Psychological Science. Dr. Elias's research, teaching, and mentoring settings include multiple universities with medical and nursing schools, a hospital, and the National Institutes of Health (NIH). Grant writing progressed from ad hoc grant reviewing, to study section memberships, to grant consultant, to center grants board membership. Significant science and grant management experience includes chartering the Adult Development and Psychopathology study section of the NIH Center for Scientific Review (CSR), serving as program officer for the National Institute on Aging, and serving as director/manager of the School of Medicine's Grants Facilitation Unit at the University of California, Davis. The multiple roles and research settings each supplied a different perspective from which to view the influence of research training, publication, grant writing, and funding on career development. These cumulative experiences provide the background and motivation for a multiple-perspective, practical guide to grant writing designed to increase the reader's potential to understand and receive funding with grant proposals.